Challenges of Human and Climate Change

ABOUT THE BOOK

Climate change has the probable to undermine sustainable development, increase poverty, and delay or prevent the realization of the Millennium Development Goals. An effective way to address the impacts of climate change is by integrating adaptation measures into sustainable development strategies so as to reduce the pressure on natural resources, improve environmental risk management, and increase the social well-being of the poor. Human rights traditionally made a distinction between state and private action. The lines are becoming increasingly blurred as a great deal of "public" power is mirrored in the global scale and enormous power of the private sector to such a degree that it is appropriate to consider the more powerful corporate actors as wielding a form of quasi-state power in relation to which it is necessary to apply human rights responsibilities to those quasi-state actors. In any event, law is one of the fundamental areas through which state power is exercised and for which states should be held accountable for serious deviancy or failure to use state power to act appropriately or to protect citizens from abuses. As the book addresses this crucial issue quite deftly, it is hoped that it would prove to be a source of great information for the reader. It is expected that the book would serve the interests of the students as a source of information for understanding and appreciating the finer aspects of Challenges of Climate Change and Human Rights.

ABOUT THE AUTHOR

Umesh Kumar, M.A, Geography is widely acclaimed name in this sphere. He has edited about ten books and published several articles in his academic fields. He is teaching from last 18 years and presently he is Asst. Professor in J.P. University, Sadhapur, Chhapra. His Dynamism is also seen in other diverse fields other than his successful academic career.

Challenges of Human and Climate Change

Umesh Kumar

AMIGA PRESS. INC
Delhi -110051

AMIGA PRESS INC.
H.O :- 28-A, Indra Park,
Chander Nagar, Delhi :- 110051 (India)
Tel :- 011-22504028, +91-98736-57000
Email :- amigapress@gmail.com
amigaeditorial@gmail.com
amigapub@gmail.com
Website :- www.amigapress.co.in

This edition is published by Amiga Press Inc.

ISBN : 978-93-87295-42-1

First Published 2019

Printed and bound by Replika Press Pvt. Ltd.

Preface

Climate change can influence humans directly, through impacts on health and the risk of extreme events on lives, livelihoods and human settlements, and indirectly, through impacts on food security and the viability of natural resource-based economic activity. The workshops and meeting discussed the impacts of climate change on achievement of the Millennium Goals in the different. Opposition for scarce resources, such as fresh water, land or fishing grounds, brought about by changes in climate, has the added potential to cause conflict over resources with impacts on the achievement of the Millennium Development Goals, and on human migration. For example, in Africa increased pressure on resources related to food and water insecurity can deepen tensions between communities and ethnic groups resulting in violence and war.

Climate change has the probable to undermine sustainable development, increase poverty, and delay or prevent the realization of the Millennium Development Goals. An effective way to address the impacts of climate change is by integrating adaptation measures into sustainable development strategies so as to reduce the pressure on natural resources, improve environmental risk management, and increase the social well-being of the poor.

Climate change is a catalyst for rising costs for human health, the global economy and the Earth's life support system. Disaster losses could reach over USD 1 trillion in a single year by 2040. Climate change presents challenges and even opportunities for the finance sector. Businesses will be affected by climate change and by policies to address it. From the perspective of sustainable development, insurance measures can be beneficial for many developing countries by transferring risk from climate change. At the workshops and expert meeting, one part of the finance sector, that of insurance, emerged as a high priority for developing countries in adapting to climate change. The SIDS expert meeting focussed particularly on insurance options.

The relationship between the development of science and technology and human rights has already been on the agenda of various United Nations' bodies for two decades. Different aspects of the probl-matique have been studied and many reports produced. The purpose of this paper is not to trace the history of the United Nations' work on the questions or to dwell at length on the human rights probl-matique, a field in which a rather rich literature exists. Its aim is to shed more light on the problem, primarily from the point of view of international law and the environment, having in mind the main goal of the study undertaken by the United Nations University in accordance with the invitation of the UN Commission on Human Rights.

This is a reference book. All the matter is just compiled and edited in nature, taken from the various sources which are in public domain.

As the book addresses this crucial issue quite deftly, it is hoped that it would prove to be a source of great information for the reader. It is expected that the book would serve the interests of the students as a source of information for understanding and appreciating the finer aspects of Challenges of Climate Change and Human Rights.

—*Editor*

Contents

1

Introduction

INSURANCE CATALYST FOR CLIMATE CHANGE

Climate change is a catalyst for rising costs for human health, the global economy and the Earth's life support system. Disaster losses could reach over USD 1 trillion in a single year by 2040. Climate change presents challenges and even opportunities for the finance sector. Businesses will be affected by climate change and by policies to address it. From the perspective of sustainable development, insurance measures can be beneficial for many developing countries by transferring risk from climate change. At the workshops and expert meeting, one part of the finance sector, that of insurance, emerged as a high priority for developing countries in adapting to climate change. The SIDS expert meeting focussed particularly on insurance options.

According to the Convention (Article 4.8), insurance-related actions constitute one of the three main means of response to the adverse effects of climate change, alongside funding and technology transfer. They can enhance financial resilience to external shocks and provide a unique opportunity to spread and transfer risk. They may provide incentives for risk reduction and prevention while engaging the private sector in climate change response action.

One of the benefits of promoting insurance-related actions is that it may help advance efforts on quantifying risks and potential losses due to climate change. Minimizing risk can result in a reduction of the rates for insurance, which thereby become more affordable. Rates could also reflect mitigation measures, such as implementation of hazard plans, forecasting and warning systems, undertaken by a community, individuals, governments, and other stakeholders.

As a result of climate change, there are both major challenges and opportunities for the insurance industry. The financial sector is already incurring additional costs from adverse climatic conditions, for example economic losses from extreme events are scaling new records. The Insurance Working Group of the UNEP Finance Initiative (UNEP FI) is a strategic and successful public-private partnership between UNEP and 16 leaders of the global financial sector. UNEP FI believe that the critical questions are "How can insurance assist developing countries grow more sustainably?" and "What are the principles for sustainable insurance?". Their recent report explores the growing area of sustainable insurance (UNEP FI 2007). At present, the insurance market is very limited in developing countries, although it is a vital instrument for these countries, particularly for coastal communities and sectors such as tourism. Current limitations are, in part, due to limited economic assets and limited private sector interest in insurance. In SIDS this is compounded by their geographical size, relative isolation, and high risk of extreme weather events.

Participants at the workshops and SIDS expert meeting highlighted the need for the implementation of insurance schemes and finance services for risk sharing at different levels. Possible cost-effective insurance initiatives. A well-coordinated dialogue between the private sector and representatives from Parties would assess cooperative actions that could be carried out to increase the insurance coverage of populations affected by climate change. The banking sector must also be involved in the work relating to climate change insurance and mainstreaming adaptation, on the basis that many loans could be at risk because of the absence of climate-proofing in projects.

In reply to the growing realization that insurance solutions can play a role in adaptation, as suggested in the UNFCCC and the Kyoto Protocol, the Munich Climate Insurance Initiative was initiated by Munich Re in April 2005.36 This initiative, formed by insurers, climate change and adaptation experts, NGOs, and policy researchers aims to develop insurance-related solutions to climate change, including identifying and promoting loss reduction measures, in cooperation with other organizations and initiatives and conduct pilot projects.

Examples of insurance schemes in developing countries from which lessons could be learned includes a microinsurance scheme by the United Insurance Company Limited Hurricane Mitigation Programme, which operates in 14 Caribbean countries. This Programme aims to reduce the vulnerability of Caribbean property to hurricanes by providing financial incentives for insurance holders to put preventative measures in place. Microfinancing is also an option for hedging risk. In Bangladesh the microfinancing institutions, Proshika and Grameen, with their long acquaintance with the impacts of disasters on the poor, have started to promote loans to reduce vulnerability to climate change. Loans are available for safer housing, diversifying incomes, from agriculture and sharecropping to more disaster-proof activities and mobile assets, and for rapid credit to promote fast recovery immediately after a disaster. Loan officers and borrowers are also increasingly taking a role in community preparedness projects.

The case of the Caribbean Catastrophe Risk Insurance Facility (CCRIF) could be extrapolated and localized to the circumstances of other regions. It uses a portion of donor-funded capital reserves to assist in the establishment of a facility that assists countries in pooling their risk and reducing insurance costs. The CCRIF uses parametric cover which enables immediate claims payments to the country because payouts are linked to triggers, such as wind speed. Insurance premiums are tied to the risk profiles of individual countries. The advantages of this scheme include efficient risk transfer mechanisms, optimal pricing from reinsurance through risk-pooling and economies of scale, and sharing of administrative and operational costs of the insurance business. However, as risk profiles increase due to the impacts of climate change, premiums will also rise which will mean that small island States will need to bear the costs of additional risks associated with climate change.

Intended for many insurance options, national governments will need to support local governments through transfer of resources based on risk assessments by subregion. Public-private partnerships with financial institutions, that help promote preparedness and mitigation and short-term training programmes for community-based organizations, could be of great assistance in building capacity at a local level. Nairobi work programme can also provide a valuable opportunity for furthering methodological efforts relating to insurance in the context of climate change adaptation. The international community could provide support in the context of insurance in a number of ways.

At the SIDS meeting, suggested ways forward include identifying specific issues and constraints relating to insurance, and engaging the insurance industry and finance experts on novel and innovative approaches to address insurance and relief funding in the context of risks relating to climate change. This could be done through expert meetings and/or workshops, perhaps bringing together actual practitioners and providers of insurance services with climate change stakeholders to devise appropriate responses to enhance the role of insurance as an adaptation tool for SIDS. This will require the involvement of non-SIDS countries to ensure practical risk distribution. Participants at the workshops and expert meeting suggested that the UNFCCC process could provide support to help identify possible insurance options and increase the resilience of countries to climate change. More information and assessment, including the expansion of early warning systems and information dissemination systems, and improvement in forecasting and disaster related decision-making would help to evaluate insurance options. The Nairobi work programme can also provide a valuable opportunity for furthering methodological efforts relating to insurance in the context of climate change adaptation.

FUNDING FOR ADAPTATION

Funding is vital in order for developing countries to plan for and implement adaptation plans and projects. A basic conclusion of the Stern Review is that the costs of strong and urgent action on climate change will be less than the costs thereby avoided of the impacts of climate change under business as usual (Stern 2006). All countries, rich and poor, need to adapt to climate change, and this will be costly. Developing countries, already the hardest hit by climate change, have little capacity (both in terms of human capacity and financial resources) to adapt. Financial ways and means must be found to enable developing countries to enhance their efforts to adapt. At the workshops and meeting the lack of funding available for adaptation was highlighted as a large impediment to implementing adaptation plans. Accessing the funds which are available at present was identified as complex and lengthy. Even if this process were to be streamlined, a lot more funding would still be required for adaptation. New international financial mechanisms to provide a sustained and sufficient response to adaptation are needed.

At the workshops and meeting, participants emphasised the need for sustained financing for adaptation. Without sustained funding, adaptation runs the risk of not being effectively addressed, and largely limited to "reactive" funding, such as short-term emergency relief. This would be unsupportive of sustainable development and ultimately prove to be very costly. Funding is required for all developing countries to develop national adaptation strategies or action plans. These plans should exist at all levels: local, provincial and national.

Article 4 of the Convention highlights that developed country Parties shall provide financial resources to assist developing country Parties adapt to climate change. To facilitate this, the Convention assigned to the GEF the responsibility of operating its financial mechanism. The GEF enables a transfer of financial resources from developed to developing countries by establishing operational programmes, providing programming documents and allocating resources. Based on guidance from the UNFCCC, the GEF operates three funds. These are the GEF Trust Fund; the Least Developed Countries Fund (LDCF) and the Special Climate Change Fund (SCCF). Further funding opportunities currently available for developing countries to fund adaptation projects include: the future Adaptation Fund under the Kyoto Protocol, funds from other multilateral environmental agreements (MEAs), and bilateral and multilateral funding from governments, national and international organizations and agencies.

The GEF Trust Fund and its Strategic Priority on Adaptation (SPA) support enabling activities and pilot and demonstration projects that address adaptation and at the same time generate global environmental benefits. COP guidance on GEF support for adaptation identified three stages. Stage I provided support for the national communications process, a portion of which is the vulnerability and adaptation assessment. Stage II provides further assistance for other capacity-building efforts for adaptation. Stage III refers to support for actual adaptation activities, including insurance. In 2001, the GEF established the Strategic Priority on Adaptation (SPA). The GEF has allocated USD 50 million under SPA of which USD 5 million has been devoted to piloting community adaptation initiatives through the Small Grants Programme (SGP). For example, the Special Program on Adaptation to Climate

Change in coastal areas aims to support participating countries (Dominica, St. Lucia and St. Vincent and the Grenadines) in implementing pilot adaptation measures, where climate change mainstreaming activities have already occurred. Activities will specifically address the impacts of climate change on biodiversity and land degradation at the coast. The Community based adaptation programme is aimed at developing a framework that spans all levels from local to intergovernmental to respond to unique community-based adaptation needs; identify and finance diverse community-based adaptation projects in a number of selected countries (Bangladesh, Bolivia, Niger, Samoa, Guatemala, Jamaica, Kazakhstan, Morocco, Namibia, Viet Nam); and disseminate lessons learned at the community level to all stakeholders, including governments.

The Special Climate Change Fund (SCCF) is partly designed to finance adaptation activities that increase resilience to the impacts of climate change, through a focus on adaptation responses particularly in water resources, land, agriculture, health, infrastructure development, disaster preparedness, and in fragile ecosystems and coastal zones. Funding will be available for establishing pilot or demonstration projects to show how adaptation planning and assessment can be practically translated into projects that will provide real benefits, and may be integrated into national policy and sustainable development planning, on the basis of information provided in the national communications, or of in-depth national studies, including NAPAs. For example, an SCCF project in Tanzania is aimed at incorporating Climate Change into integrated Water Resources Management in the Pangani River Basin; and in Ecuador an SCCF project is enabling adaptation to climate change through effective water governance.

The Least Developed Countries Fund (LDCF) was partly established to support projects addressing urgent and immediate adaptation needs in the least developed countries as identified by their National Adaptation Programmes of Action (NAPAs). The LDCF supports a learning-by-doing approach to adaptation. Projects proposed in the NAPAs which are being considered for LDCF funding include a coastal afforestation community based adaptation project in Bangladesh; a project to integrate climate change risks into community-based livestock management in the northwest lowlands of Eritrea; and a project to reduce climate change-induced risks and vulnerabilities from glacial lake outburst floods in the Punakha-Wangdi and Chamkhar Valleys in Bhutan.

At the workshops and expert meeting developing country representatives expressed their concern with the complexity of current arrangements for accessing funds from the GEF. It was highlighted that procedures for accessing existing financial resources for adaptation remain complex and lengthy. Participants at the UNFCCC workshops and meeting emphasised the need for capacity-building and information on the financial mechanisms available to improve access to funding and to facilitate project preparation. Improving the access of developing countries to financial resources, including through streamlining guidelines for application and by assisting countries in the preparation of project proposals was thought important at all workshops and meeting.

Within the context of the UNFCCC, the international carbon market has emerged as a result of the Clean Development Mechanism established under the Kyoto Protocol. The Clean Development Mechanism allows industrialized countries to help generate funding for adaptation in developing countries in the context of sustainable development while providing them a cost-effective means of offsetting their greenhouse gas emissions. The mechanism enables approved emission-reducing projects in developing countries to earn certified emission reduction units, each equivalent to one tonne of carbon dioxide, which the project participants in the developing country can then sell to buyers in industrialized countries. These sustainable development projects range from wind farms to hydroelectric power stations and also include energy efficiency projects.

The Adaptation Fund under the Kyoto Protocol is intended to fund concrete adaptation projects and programmes in developing countries that are particularly vulnerable to the adverse effects of climate change. The source of this funding is intended to be from a 2 per cent levy on proceeds from Clean Development Mechanism projects (excluding those undertaken in least developed countries), as well as from other voluntary sources. The Adaptation Fund is yet to be operationalized. The actual amount of money that will be available from this fund is uncertain as it depends on the extent of use of the Clean Development Mechanism and on the price of carbon. The World Bank estimates that the Adaptation Fund is likely to total USD 100 – 500 million by 2012.

Multilateral Environmental Agreements (MEAs) whose areas of work could be synergetic with adaptation may also provide further funding for adaptation. These MEAs include the Convention on Biological Diversity, UN Convention to Combat Desertification and the Ramsar Convention on Wetlands. For example, the Central African Regional Program for the Environment funded by USAID aims to help mitigate climate change by absorbing and storing carbon dioxide from the atmosphere, promoting biodiversity conservation and improving forest management, and sustainable agriculture, and help reduce the vulnerability of ecosystems.

Bilateral and Multilateral Funding includes funding of adaptation projects directly via inter-governmental organizations, national and local governments, institutions and non-governmental organizations. Examples include many community-based projects by non-governmental organizations aimed at improving agricultural practices, water resource supply and use, primary health care and energy supply in developing countries.

In 2007 the UNFCCC secretariat launched a project to review and analyse investment and financial flows relevant to the development of an effective and appropriate international response to climate change, with particular focus on developing country's needs, including their medium-to long-term requirements for investment and finance.

As part of this project, the secretariat assessed the investment flows needed in 2030 that will be necessary to meet adaptation requirements in several sectors. The global cost of adaptation to climate change is difficult to estimate as climate change adaptation measures will be widespread and heterogeneous. However, for all of the sectors examined, there is a substantial deficit in current investment and financial flows.

In 2030 the adaptation funds required were estimated at: USD 14 million for agriculture, forestry and fisheries; USD 11 billion for water resources; USD 5 billion for human health; USD 11 billion for coastal zones; and USD 8 –130 billion for infrastructure. In summary, the UNFCCC secretariat estimated that the investment and financial flows needed for adaptation are likely to be tens of billions of dollars per year several decades from now and could be more than USD 100 billion per year. Other studies (World Bank 2006b, Oxfam 2007) also estimate adaptation costs at tens of billions of dollars per year.

The funds that are currently available under the Convention and the Kyoto Protocol are small compared to the magnitude of the needs identified by the UNFCCC. The financial resources available for adaptation in the funds currently operated by the GEF (Trust Fund, LDCF and SCCF) amounted to about USD 275 million as of August 2007. The Adaptation Fund could receive USD 80 – 300 million per year for the period 2008 – 2012. Assuming a share of proceeds for adaptation of 2 per cent continues to apply post 2012, the level of funding could be: USD 100 – 500 million per year for a low demand for the CDM; and USD 1– 5 billion per year for a high demand. However, there is still a deficit in funding that needs to be filled.

INFORMATION ANALYSIS - REPORTING OF IMPACTS, VULNERABILITY AND ADAPTATION

Reliable, systematic climate data helps countries determine their current climate variabilities, and model future changes. Countries use a number of assessment models, tools and methodologies as well as various scenarios, including those provided by the IPCC (IPCC 2000), to help provide an assessment of the future impacts of climate change. Climate change impacts, vulnerability and adaptation assessments need to generate outputs that are policy relevant. To do this, climate change data including future impacts and vulnerabilities needs to be integrated with socioeconomic data and analyses across a range of sectors, and the results must be tailored for policymakers and stakeholders.

All Parties to the UNFCCC are committed to submit national communications in which they outline the implementation of the UNFCCC and the impacts from climate change that they are facing. In their national communications, countries provide an assessment of vulnerabilities and adaptation options. Water resources, agriculture and food security, human health, terrestrial ecosystems and biodiversity, and coastal zones are common sectors for which impacts and vulnerability assessment have been carried out by developing countries.

At the workshops and expert meeting, vulnerability and adaptation assessments were identified as vital tools for developing countries to evaluate and implement responses to climate change. A major problem in all regions was the limited capacity at regional and national level due to deficiencies in data collection and the lack of technical expertise. It was highlighted as important to make the models, tools and methodologies that are appropriate for assessments in developing countries more widely available. Exchanging information on tools used for vulnerability and adaptation assessments, together with the outcomes of these assessments, would help countries improve capacity in this area. This could be done through workshops and symposia, regional science journals, websites to facilitate information exchange and by making better use of existing channels of information.

The resolution of models used to determine climate change in developing countries is too course and often relies on data from sources in other countries. Along with the disparity in outputs from different models, this makes the use of results as a basis for adaptation action very difficult. A major problem encountered when using models to get national results is the need for readjustment and downscaling to suit a country's individual needs. Participants highlighted the need for enabling training on the use of models and tools in all regions and providing technical support, such as through the training of trainers on modelling tools, and follow-up through regional centres of excellence, to address gaps in expertise in the application of methods and tools and in using climate information and geographical information systems. At the SIDS meeting special mechanisms for vulnerability and adaptation assessment training, including short courses and longer-term professional training incorporating capacity-building for participatory approaches, were mentioned as a follow-up action.

The development of higher resolution regional models for developing countries is important as well as analysing the disparity between the model outcomes. This would help enhance capacity for reaching informed decision making. For example, at the Africa workshop, participants emphasised the need to develop regional climate models to provide fine-scale climate information for long-term impact studies and forecasting, as well as facilitate information exchange between African institutions. Some efforts are being undertaken in this regard in developing countries, and regional models are being developed that are capable of providing more useful information needed by planners and policy makers. For example, the Hadley Centre's model PRECIS (Providing Regional Climates for Impacts Studies) has been designed for use by local meteorological offices or research institutes. Training on this model has been undertaken in several developing countries, including Cuba, Brazil, Argentina, South Africa and India, and Jamaica, Cuba and Barbados mentioned at the expert meeting that results from the model have been used in vulnerability assessments.

At the SIDS meeting it was further highlighted that establishing a group of experts to facilitate assessments for specific circumstances of SIDS would be useful. Establishing a roster of experts with specialized skills, in all regions at centres of excellence, would help to maintain experts working in the regions. This would also help continuity in impact and vulnerability assessment, a problem highlighted at the Africa meeting where participants emphasized that the impact and vulnerability assessment undertaken in the initial national communication process was disjoined from that of the second national communication.

Participants at all workshops and at the expert meeting emphasised the lack of socio-economic data, or indeed development indicators and relevant tools for enhancing the use of socio-economic assessment in a way that would be relevant to policy makers and other stakeholders, including ministries of finance and economy. It is important to link climate vulnerability to socio-economic studies and long-term periodic and socio-economic assessments. Preserving indigenous knowledge that is relevant to community level responses, studies on coping strategies, and gender specific vulnerability assessments were all highlighted as important elements to determining adaptation options.

Useful methodologies for assessing adaptation options include both top-down and bottom-up approaches. Both methodologies need to be linked to promote integrated adaptation assessments. Top-down methodologies include the use of modelling and scenario analysis. This can provide useful background to decision making and is strong in terms of the biophysical aspects of impacts. However the models do not perform well in representing human interactions and local abilities to adapt. This is highlighted by an example from the Cook Islands (Box III- 2).

As a complement to the top-down approach is a vulnerability-based, bottom-up, approach, which recognizes and builds upon local coping strategies and indigenous knowledge and technologies, and the capacity and coping range of communities, local institutions and sectors in responding to current climate variability. This approach helps to incorporate human and economic dimensions of the local communities, particularly livelihood aspects and inter-sectoral relationships. It is useful in developing specific strategies and policy implementation. However, it exhibits a weaker attribution to future climate change.

An example of this approach is the UNFCCC's National Adaptation Programmes of Action (NAPAs) for use by least developed countries to prioritize their urgent adaptation needs. The rationale for NAPAs rests on the limited ability of least developed countries to assess their vulnerability and adapt to climate change. A new approach was needed that would focus on enhancing adaptive capacity to climate variability and thus help these countries directly address their urgent needs arising from the adverse effects of climate change. The NAPAs use and build upon existing coping strategies at the grassroots level, rather than focusing on scenario-based modelling, to assess future vulnerability and adaptive responses at local and state level. Involvement of different stakeholders (national, sectoral, local) and including existing coping strategies are an integral part in the assessment process.

EFFECTS OF SYSTEMATIC CLIMATE VARIABILITY ON EPIDEMICS

Another significant climate trend that has been linked to systematic changes in temperature and precipitation is an increase in the variability, or extremes, of climate. This change in extremes can alter not only the intensity of individual events, such as storms and floods, but the timing and spatial patterns of weather as well. Since the mid-1800s the average surface temperature of the globe has risen about 0.4°C-0.6°C, and periods of persistent warming can, in general, be associated with increased variability. The IPCC projects that the warming trend may be accompanied by more intense heat waves and altered drought and rainfall patterns. As reported in the first issue of Consequences, data from the National Climatic Data Centre—this nation's main repository for meteorological data—indicate that, since the 1970s, extreme weather events have indeed increased in the continental U.S. On average, periods of drought are systematically longer, and bursts of precipitation (greater than two inches of rain over twenty-four hours) are more frequent. A warmed atmosphere accelerates evaporation, and for every rise of 1°C, the air can hold 6 percent more water. One consequence is that we are now receiving a greater percentage of precipitation in the form of sudden, intense bursts that are more typical, for example, in the tropics. Longer droughts and more heavy bursts of rain, accompanied by flash floods, were more common in the 1980s than in the 1970s, and more so in the 1990s than in the 1980s.

Extreme events—floods, storms, droughts, and uncontained fires—can be devastating for agriculture, for human settlements, and for health. More, brief cold snaps are also possible, and winter storms, like heat waves, bring an increase in cardiac deaths. Floods spread bacteria, viruses, and chemical contaminants, foster the growth of fungi, and contribute to the breeding of insects. Prolonged droughts, interrupted by heavy rains, favour population explosions of both insects and rodents. Extreme weather events (often associated with the recurring climatic conditions that are initiated by large-scale changes in sea surface temperatures in the Pacific, known as El Nino/La Nina events) have been accompanied by new appearances of harmful algal blooms in Asia and North America, and—in Latin America and Asia—by outbreaks of malaria and various water-borne diseases, such as typhoid, hepatitis A, bacillary dysentery, and cholera.

Rodents: Synergies and Surprises

Rodents are today a growing problem in the U.S., Latin America, Africa, Europe, Asia, and Australia. These preeminent opportunists are believed to be the fastest reproducing mammal and they eat everything humans do, thrive on contaminated water and food, and are extremely capable swimmers. Meadow voles, whose numbers are kept in check by predatory marsh hawks, can have up to seventeen broods a year, for example, each of half a dozen offspring. Rodents consume 20 percent of the world's grain, including almost a seventh in our own country, and up to three quarters of what is grown and stored in some African nations. Rodents can also carry diseases. A controlled experiment with Canadian snowshoe hares depicts how multiple factors can act synergistically to greatly increase the number of rodents. Excluding predators by confining the rabbits to cages led to a population doubling, compared to the fate of a number of similar animals in the wild. Augmenting food tripled hare density. Together the interventions in the controlled experiment resulted in more than a ten-fold population explosion.

Rats, Mice, and the Hantavirus

The story of the hantavirus illustrates a similar synergy in the case of microbial agents. A prolonged drought in the U.S. southwest in the early 1990s reduced the populations of animals such as owls, coyotes, and snakes that prey on rodents. When the drought yielded to intense rains in 1993, the grasshoppers and pinon nuts on which rodents feed became more abundant. The result, when combined with the drop in predators, was a ten-fold increase in rats and mice by June of that year. An outcome was the emergence of a "new" disease called Hantavirus Pulmonary Syndrome: from a virus—perhaps already present, but dormant—transmitted through rodent saliva, urine, and droppings. Predators returned, however, and by summer's end the outbreak had abated. Rodent-borne hantaviruses have resurged in several European nations, and most notably in the former Soviet Union and in Yugoslavia. In late 1996, hantavirus infection emerged in South America in western Argentina. Within the first few months at least ten deaths resulted, frightening off tourists and threatening the economic livelihood of the region. Hantavirus Pulmonary Syndrome has now appeared in Bolivia, Brazil, Canada, Chile, Ecuador, Paraguay, and Uruguay, with about 50 percent mortality and several cases of person-to-person transmission.

Disease Clusters

In August 1995 the eastern, tropical region of the Pacific Ocean surface turned cold, initiating a La Nina event that would last until late 1996. Along the Caribbean coast of Colombia a summer 1995 heat wave was followed by the heaviest August rainfall in fifty years, ending a long drought that accompanied the preceding, prolonged El Nino conditions of 1990-1995. The heat and flooding precipitated a cluster of diseases involving mosquitoes (Venezuelan equine encephalitis and dengue fever), rodents (leptospirosis), and toxic algae (that killed 350 tons of fish in their largest coastal lagoon). Prolonged anomalous conditions of the sort that applied in 1990-1995 can also have cumulative biological consequences. In New Orleans, for example, five years without a killing frost was associated with an explosion of mosquitoes, termites, and cockroaches. Termites persisted inside trees into the cold winter of 1995-1996, and now threaten to destroy stands of New Orleans' fabled "mighty oaks." We may have only just begun to understand the true biological impacts of the persistent anomalous climate of the 1990s.

Some other Examples

Evidence of another rodent-borne disease —leptospirosis— is increasingly reported in U.S. urban centres, in areas where the disposal of sewage and other measures to protect public health have declined. In 1995, there were substantial outbreaks of the disease in Central America and Colombia, as heavy (La Nina) rains, following the prolonged El Nino drought, drove rodents scurrying from their burrows. Leptospirosis is treatable with antibiotics; but there were fatalities in 1995 before the diagnosis was established.

A combination of stresses contributed to the sudden appearances of several rodent-borne viral hemorrhagic fevers in rural Latin America in the past several decades: Junin in Argentina (1953), machupo in Bolivia (1962 and 1996), guaranito in Venezuela, and sabia in Brazil. In Bolivia, systematic clearing of trees apparently shifted populations of a variety of disease-carrying mice, known as calomys, from forest to field settings where they became dominant. Heavy applications of DDT —meant to eradicate malaria— helped reduce their natural predators. When cats were reintroduced to the area in 1962, the epidemic of Bolivian hemorrhagic fever was abated, although not until it had killed 10 to 20 percent of the inhabitants of the small villages in which the disease was present. Habitat loss and excessive use of toxic chemicals acted together in this case. In southern Africa, rodent populations exploded as a consequence of climate variability, when heavy and then lighter rains came in 1993 and 1994, on the heels of six years of prolonged drought. When the rains came, rodents found themselves in a world where avian and land predators were virtually absent.

Moreover, because so many draft animals had also succumbed, there was little tillage of the land and the underground burrows in which rodents live went largely undisturbed. After an initial successful harvest in 1993, the maize crop in Zimbabwe was decimated by rodents. Soon after, human plague broke out in Zimbabwe and on the borders of neighbouring Malawi and Mozambique, carried—as in the devastating plague of 14th century Europe—by fleas on rats. Subsequently, a rodent-borne virus took the lives of eighty-one elephants in South Africa's Kruger Park, and plague returned in the summer of 1997. Plague in India resurfaced in 1994, following a blistering summer when temperatures reached 124°F, leaving animals prostrate in the northern part of the country and fueling the breeding of fleas in houses that held stored grain. The unusually heavy monsoons following the 1994 heat wave led to population crowding in Surat on the western coast, north of Bombay, and an apparent outbreak of pneumonic (person-to-person) plague. Cases of malaria and dengue fever also surged in the wake of flooding. Meanwhile, in Australia, rodents emerged as serious crop pests in 1995, accompanying the same prolonged El Nino of the early 1990s and the ensuing years of intense drought.

Current land-use practices and the overuse of chemicals to control pests may increase the chances for such "nasty synergies." Climate variability is also a key element in upsurges of pests—and were climate to become more unstable, it could exert an even greater influence on ecological dynamics and the patterns of infectious diseases in the future. A disturbance in one factor can be destabilizing; but multiple perturbations can reduce the resistance and the resilience of a entire system.

Loss of Biological Diversity

The extent of biological impoverishment all over the globe has been a source of great concern for many years. If the present trend in destruction of biotopes continues, a quarter of all plant species, along with the fauna that depend on them, will disappear within the next century. Human activities and the spread of human beings, as a result of population growth, into hitherto untouched ecosystems are the main reason for the reduction of the biological diversity.

The total number of species on earth is not known. Only about 1.4 million species have been identified, but estimates range from five to thirty million. Today, this natural variety is more profoundly threatened than ever before in the last 65 million years. While human beings only see the larger animals and plants and worry about there extinction, 95 percent of species are lost almost without being noticed. This loss is regrettable for esthetic, ethical, philosophical and ecological reasons.

The loss of biodiversity is also deplorable from an economic standpoint since it involves the irrecoverable loss of genetic resources whose value for man and nature has never even been determined. One can only speculate whether tropical rain forests contain basic materials for decisive breakthroughs in the treatment of diseases, which today are still regarded as incurable. To date, less than a tenth of one percent (<0.1%) of naturally occurring species have been used.

The present state of our knowledge does not permit us to judge the indirect value of biodiversity emerging from the varied (and still largely unknown) reciprocal relationships between animals, plants and microorganisms as they affect the various eco-systems. According to various estimates, as much as 50 to 75 percent perhaps even 90 percent, of all species are native to the tropical rain forests (which constitute only 7 percent of the earth's surface). Since these forests are currently being destroyed more rapidly than all other habitats on earth, the extinction of species is today greater than ever before in the history of mankind - presumably one species per hour since the mid-eighties. Of the 45,000 to 50,000 species of plants in tropical Asia alone, 300 are disappearing forever each year.

Other environmental problems such as lack of fuel wood, inadequate sanitation or water scarcity are exacerbated by population growth. Most of the ecological consequences connected with rapid population growth are of a national and regional nature - some, like the destruction of forests on a large scale, have also a global impact. Although the "rich" minority of this world is in ecological terms clearly the more destructive element than the "poor" majority, endeavors to lower population growth are a vital precondition for sustainable development.

Terminology

The most general definition of *climate change* is a change in the statistical properties of the climate system when considered over periods of decades or longer, regardless of cause. Accordingly, fluctuations on periods shorter than a few decades, such as El Nino, do not represent climate change.

The term sometimes is used to refer specifically to climate change caused by human activity; for example, the United Nations Framework Convention on Climate Change defines climate change as "a change of climate which is attributed directly or indirectly to human activity that alters the composition of the global atmosphere and which is in addition to natural climate variability observed over comparable time periods." In the latter sense climate change is synonymous with global warming.

Causes

Factors that can shape climate are climate forcings. These include such processes as variations in solar radiation, deviations in the Earth's orbit, mountain-building and continental drift, and changes in greenhouse gas concentrations. There are a variety of climate change feedbacks that can either amplify or diminish the initial forcing. Some parts of the climate system, such as the oceans and ice caps, respond slowly in reaction to climate forcing because of their large mass. Therefore, the climate system can take centuries or longer to fully respond to new external forcings.

Plate Tectonics

Over the course of millions of years, the motion of tectonic plates reconfigures global land and ocean areas and generates topography. This can affect both global and local patterns of climate and atmosphere-ocean circulation.

The position of the continents determines the geometry of the oceans and therefore influences patterns of ocean circulation. The locations of the seas are important in controlling the transfer of heat and moisture across the globe, and therefore, in determining global climate.

A recent example of tectonic control on ocean circulation is the formation of the Isthmus of Panama about 5 million years ago, which shut off direct mixing between the Atlantic and Pacific Oceans. This strongly affected the ocean dynamics of what is now the Gulf Stream and may have led to Northern Hemisphere ice cover. During the Carboniferous period, about 300 to 360 million years ago, plate tectonics may have triggered large-scale storage of carbon and increased glaciation. Geologic evidence points to a "megamonsoonal" circulation pattern during the time of the supercontinent Pangaea, and climate modeling suggests that the existence of the supercontinent was conducive to the establishment of monsoons. The size of continents is also important. Because of the stabilizing effect of the oceans on temperature, yearly temperature variations are generally lower in coastal areas than they are inland. A larger supercontinent will therefore have more area in which climate is strongly seasonal than will several smaller continents or islands.

Solar Output

Variations in solar activity during the last several centuries based on observations of sunspots and beryllium isotopes. The sun is the predominant source for energy input to the Earth. Both long-and short-term variations in solar intensity are known to affect global climate.

Three to four billion years ago the sun emitted only 70% as much power as it does today. If the atmospheric composition had been the same as today, liquid water should not have existed on Earth. However, there is evidence for the presence of water on the early Earth, in the Hadean and Archean eons, leading to what is known as the faint young sun paradox. Hypothesized solutions to this paradox include a vastly different atmosphere, with much higher concentrations of greenhouse gases than currently exist Over the following approximately 4 billion years, the energy output of the sun increased and atmospheric composition changed, with the oxygenation of the atmosphere around 2.4 billion years ago being the most notable alteration. These changes in luminosity, and the sun's ultimate death as it becomes a red giant and then a white dwarf, will have large effects on climate, with the red giant phase possibly ending life on Earth.

Solar output also varies on shorter time scales, including the 11-year solar cycle and longer-term modulations. Solar intensity variations are considered to have been influential in triggering the Little Ice Age, and some of the warming observed from 1900 to 1950. The cyclical nature of the sun's energy output is not yet fully understood; it differs from the very slow change that is happening within the sun as it ages and evolves. While most research indicates solar variability has induced a small cooling effect from 1750 to the present, a few studies point toward solar radiation increases from cyclical sunspot activity affecting global warming.

Orbital Variations

Slight variations in Earth's orbit lead to changes in the seasonal distribution of sunlight reaching the Earth's surface and how it is distributed across the globe. There is very little change to the area-averaged annually averaged sunshine; but there can be strong changes in the geographical and seasonal distribution. The three types of orbital variations are variations in Earth's eccentricity, changes in the tilt angle of Earth's axis of rotation, and precession of Earth's axis. Combined together, these produce Milankovitch cycles which have a large impact on climate and are notable for their correlation to glacial and interglacial periods, their correlation with the advance and retreat of the Sahara, and for their appearance in the stratigraphic record.

Volcanism

Volcanism is a process of conveying material from the crust and mantle of the Earth to its surface. Volcanic eruptions, geysers, and hot springs, are examples of volcanic processes which release gases and/or particulates into the atmosphere. Eruptions large enough to affect climate occur on average several times per century, and cause cooling (by partially blocking the transmission of solar radiation to the Earth's surface) for a period of a few years.

The eruption of Mount Pinatubo in 1991, the second largest terrestrial eruption of the 20th century (after the 1912 eruption of Novarupta) affected the climate substantially. Global temperatures decreased by about 0.5 °C (0.9 °F). The eruption of Mount Tambora in 1815 caused the Year Without a Summer. Much larger eruptions, known as large igneous provinces, occur only a few times every hundred million years, but may cause global warming and mass extinctions. Volcanoes are also part of the extended carbon cycle. Over very long (geological) time periods, they release carbon dioxide from the Earth's crust and mantle, counteracting the uptake by sedimentary rocks and other geological carbon dioxide sinks. According to the US Geological Survey, however, estimates are that human activities generate more than 130 times the amount of carbon dioxide emitted by volcanoes.

Ocean Variability

The ocean is a fundamental part of the climate system. Short-term fluctuations (years to a few decades) such as the El Nino–Southern Oscillation, the Pacific decadal oscillation, the North Atlantic oscillation, and the Arctic oscillation, represent climate variability rather than climate change. On longer time scales, alterations to ocean processes such as thermohaline circulation play a key role in redistributing heat by carrying out a very slow and extremely deep movement of water, and the long-term redistribution of heat in the world's oceans.

Human Influences

Anthropogenic factors are human activities that change the environment. In some cases the chain of causality of human influence on the climate is direct and unambiguous (for example, the effects of irrigation on local humidity), while in other instances it is less clear. Various hypotheses for human-induced climate change have been argued for many years. Presently the scientific consensus on climate change is that human activity is very likely the cause for the rapid increase in global average temperatures over the past several decades. Consequently, the debate has largely shifted onto ways to reduce further human impact and to find ways to adapt to change that has already occurred.

Of most concern in these anthropogenic factors is the increase in CO_2 levels due to emissions from fossil fuel combustion, followed by aerosols (particulate matter in the atmosphere) and cement manufacture. Other factors, including land use, ozone depletion, animal agriculture and deforestation, are also of concern in the roles they play-both separately and in conjunction with other factors-in affecting climate, microclimate, and measures of climate variables.

CLIMATE CHANGE SCENARIOS

Sensitivity tests: Arbitrary climate sensitivity tests were conducted to test crop model responses to a range of temperature (+2deg.C and +4deg.C) and precipitation (+/-20%) changes. While this type of sensitivity analysis is dissociated from the processes that influence climate (temperature and precipitation are physically related in any given region), it does provide better understanding of the factors affecting crop model responses.

It can also help to identify climatic thresholds of critical impacts. Sensitivity tests were carried out in 13 of the 18 countries. The sensitivity test results were not utilized in the economic analysis with the BLS world food trade model because of their lack of realism, e.g., temperature change at high latitudes is predicted to be greater than the global mean in winter (IPCC, 1990a), rather than increase by 2deg. or 4deg.C in all regions of the world throughout the year. Farm-level adaptations were not tested in the sensitivity studies.

Scenarios based on general circulation model results: Scenarios of climate change were developed in order to estimate their effects on crop yields and food trade. A climate change scenario is defined as a physically consistent set of changes in meteorological variables, based on generally accepted projections of concentrations of carbon dioxide (CO_2) and other greenhouse gases in the atmosphere (thought to be the likely cause of future climate change). The set of scenarios used is intended to capture a range of possible effects and set limits on the associated uncertainty. The scenarios for this study were created by changing observed daily data from the current climate (1951-80) according to doubled CO2 simulations of three general circulation models available at the initiation of the study in 1989. The GCMs used are those from the Goddard Institute for Space Studies (GISS), Geophysical Fluid Dynamics Laboratory (GFDL), and United Kingdom Meteorological Office (UKMO). The temperature changes of these GCM scenarios (4.0-5.2deg.C) are at or near the upper end of the range (1.5-4.5deg.C) projected for doubled CO2 warming by the IPCC (IPCC, 1990a, 1992). The GISS and GFDL scenarios, however, are near the mean temperature change (3.8deg.C) of recent doubled CO_2 experiments documented for atmospheric GCMs with a seasonal cycle and a mixed-layer ocean (IPCC, 1992). Mean monthly changes in temperature, precipitation, and solar radiation from the appropriate GCM gridbox were applied to observed daily climate records to create climate change scenarios for each site.

Scientific uncertainty of GCM predictions: While GCMs currently provide the most advanced means of predicting the potential future climatic consequences of increasing amounts of radiatively active trace gases, their ability to simulate current climate varies considerably from region to region. They have been shown to simulate current surface air temperatures reasonably well, but do not reproduce current precipitation as accurately (IPCC, 1990a). Of special importance for agricultural climate change impacts, there is a notable lack of consensus among GCMs in prediction of regional soil moisture changes (Kellogg and Zhao, 1988). Furthermore, GCMs are not yet able to produce reliable projections of changes in climate variability, such as alterations in the frequencies of drought and storms, even though these could affect crop yields significantly.

CO_2 level and timing: For the crop modeling part of this study, climate changes from doubled CO_2 GCM simulations are utilized with an associated level of 555 ppm CO_2; the assumed timing for the BLS world food trade projections is that these conditions will occur in 2060. This approach, which mixes equilibrium climate and transient CO_2 projections, is the best that can be done given the current lack of availability of GCM transient climate change simulations for impact studies and the need for a dynamic time course for the economic model. It is not known what rates of future emissions of trace gases will be and when the full magnitude of their effects will be realized.

Because atmospheric concentrations of other greenhouse gases besides CO_2 (e.g.. methane (CH4), nitrous oxide (N2O)), and the chlorofluorocarbons (CFCs)) are also increasing, an 'effective CO_2 doubling' has been defined as the combined radiative forcing of all greenhouse gases having the same forcing as doubled CO_2. Level of CO_2 is important when estimating potential impacts on crops, because crop growth and water use have been shown to benefit from increased levels of CO_2 (Cure and Acock, 1986). A CO_2 level of 555 ppm was associated with the effective doubled CO_2 climate projections for use in the crop modeling simulations. This was based on the GISS GCM transient trace gas scenario A described in Hansen *et al.* (1988), in which the simulated climate has warmed to the effective doubled CO_2 level of about 4degC by 2060. This level assumes that non-CO_2 trace gases contribute ~15% of the change in radiative forcing from 300 to ~600 ppm.

CULTURAL RIGHTS

Every human being has the right to culture, including the right to enjoy and develop cultural life and identity. Cultural rights, however, are not unlimited. The right to culture is limited at the point at which it infringes on another human right. No right can be used at the expense or destruction of another, in accordance with international law.

This means that cultural rights cannot be invoked or interpreted in such a way as to justify any act leading to the denial or violation of other human rights and fundamental freedoms. As such, claiming cultural relativism as an excuse to violate or deny human rights is an abuse of the right to culture.

There are legitimate, substantive limitations on cultural practices, even on well-entrenched traditions. For example, no culture today can legitimately claim a right to practise slavery. Despite its practice in many cultures throughout history, slavery today cannot be considered legitimate, legal, or part of a cultural legacy entitled to protection in any way. To the contrary, all forms of slavery, including contemporary slavery-like practices, are a gross violation of human rights under international law.

Similarly, cultural rights do not justify torture, murder, genocide, discrimination on grounds of sex, race, language or religion, or violation of any of the other universal human rights and fundamental freedoms established in international law. Any attempts to justify such violations on the basis of culture have no validity under international law.

Cultural Context

The argument of cultural relativism frequently includes or leads to the assertion that traditional culture is sufficient to protect human dignity, and therefore universal human rights are unnecessary. Furthermore, the argument continues, universal human rights can be intrusive and disruptive to traditional protection of human life, liberty and security.

When traditional culture does effectively provide such protection, then human rights by definition would be compatible, posing no threat to the traditional culture. As such, the traditional culture can absorb and apply human rights, and the governing State should be in a better position not only to ratify, but to effectively and fully implement, the international standards.

Traditional culture is not a substitute for human rights; it is a cultural context in which human rights must be established, integrated, promoted and protected. Human rights must be approached in a way that is meaningful and relevant in diverse cultural contexts. Rather than limit human rights to suit a given culture, why not draw on traditional cultural values to reinforce the application and relevance of universal human rights? There is an increased need to emphasize the common, core values shared by all cultures: the value of life, social order and protection from arbitrary rule. These basic values are embodied in human rights.

Traditional cultures should be approached and recognized as partners to promote greater respect for and observance of human rights. Drawing on compatible practices and common values from traditional cultures would enhance and advance human rights promotion and protection. This approach not only encourages greater tolerance, mutual respect and understanding, but also fosters more effective international cooperation for human rights.

Greater understanding of the ways in which traditional cultures protect the well-being of their people would illuminate the common foundation of human dignity on which human rights promotion and protection stand. This insight would enable human rights advocacy to assert the cultural relevance, as well as the legal obligation, of universal human rights in diverse cultural contexts. Recognition and appreciation of particular cultural contexts would serve to facilitate, rather than reduce, human rights respect and observance.

Working in this way with particular cultures inherently recognizes cultural integrity and diversity, without compromising or diluting the unquestionably universal standard of human rights. Such an approach is essential to ensure that the future will be guided above all by human rights, non-discrimination, tolerance and cultural pluralism.

LAND RIGHTS, LINGUISTIC RIGHTS, AND CULTURAL AUTONOMY

In many ways over the past few years the position of the Finnish Sami people has improved considerably. However, since early 1995, the Finnish Sami Parliament has observed with a growing concern that, at the same: time that Finland secures or intends to secure a special position and rights for the Sami population in its legislation, in administration and in the implementation of treaties, the acts and organised activities against the Sami are increasing in the Sami homelands. Some members of the majority population - and some municipalities (local governments) - do not accept the fact that the Sami people are guaranteed the use of their own language and the exercise of their own culture with arrangements that do not apply to the rest of the population. In addition, some news media and journalists transmit, without the slightest criticism, propaganda against the Sami and express their own anti-Sami attitudes.

Either because of the Penal Code or entrenched practices, public prosecutors do not *sua salute* prosecute anti-Sami acts which have already become rooted in everyday life in our society. The organised anti-Sami activities by the majority population are a source of anxiety for the Sami, frightening some of them to the extent that they no longer dare behave as Semi. The situation essentially weakens their *de facto* possibilities to use their mother tongue, practice their own culture and assume their own identity and own way of life. The anti-Sami actitivities also have a negative effect on political decision-making, creating difficulties for legislative solutions affecting the Sami population.

THE ELEVENTH AND TWELFTH PERIODIC REPORTS OF FINLAND

At its forty-eighth session, from 26 February to 15 March 1996, the UN Committee on the Elimination of Racial Discrimination in item 12 expressed its concern over the Sami people's participation in the 'Sami Parliament' in their own mother tongue; and in item 24 recommended that the State Party do all in its power to enable Sami children to persue their studies at primary and seccondary levels in their mother tongue. On item 12, the Sami Parliament would like to specify that three Sami language are used in Finland: Northern Sami, Inari Sami, and Skolt Sami. The legislature permits the use of all three languages in the Sami Parliament both for meetings and for records. The Northern Sami is used for both purposes systematically. The other two can be used f'or meetings with simultaneous interpretation. Their use is, however, undermined by a lack of resources; both financial factors and a scarcity of qualified interpreters.

As regards item 24, the Sami Parliment states that primary schools in the municipality of Utsjoki have classes where pupils are taught in the Sami language and in the municipality of Inari teaching of Sami-speaking children is proximately half and half in Sami and Finnish. Schools at secondary level particularly are afflicted by a shortage of materials in the Sami language and of Sami-speaking teachers of special subjects.

It is estimated that there are only a few children under and of school-age whose mother tongue is Inari Sami or Skolt Sami. The teaching of these languages in the schools is decreasing, because they are optional subjects. Whether or not children take optional language classes is decided by parents, who were forbidden even to speak the Sami language at school till the 1960s. These parents do not dare put their children in classes which are given in Sami for fear that the children will not learn enough Finnish. The speakers of Inari Sami and Skolt Sami both number 400 to 500 persons. These languages are dying out in Finland and in the world. The government and the Finnish scientific community show interest in the matter in too few practical measures or allocate too scarce resources for such measures. In fact, not much would be needed to turn the development round.

The problems of education in the Sami language are related isolation, school administration, and resources. Our legislation does not require with sufficient unambiguity that Sami children be taught the Sami language and be given school education in that language. Schools are administered by municipalities. The Sami population does not enjoy a specific legal position in the decision-making process on the provision of education for Sami children and in the related use of government grants. These funds are not earmarked for the Sami population. In recent years the reverse trend in government funding has led to a halt in the previous, positive development, and the municipalities have not been able to bring the education of Sami children to the level of goals specified in the law. The teaching of the Sami minority languages: Inari Sami and Skolt Sami, is at risk for yet another reason: they are optional subjects in schools.

The Sami language is the foundation of the Sami education and culture. Education in the Sami language and the teaching of that language are part of the linguistic rights of Sami people and of the basic services for citizens. The administration of basic services lor the Sami should, in accordance with the Act on Sami Cultural Autonomy, be administered by the Sami themselves, instead of being dependent on the various bodies in municipal school administration. on their composition. and on the political climate at a given moment. In spite of the legislation, the provision of daycare in the Sami language does not function at nearly a satisfactory level; in some places it is not provided at all. The Sami language and culture are not adequately transmitted by the generations of grandparents and parents to those children under school-age who do not receive daycare in that language. In the next phase. the school cuts the children off from their cultural background and its values.

If the Sami language and culture are not transmitted in daycare and school education from one generation to the next, they will die; as will happen with the minority languages of Inari Sami and Skolt during the present generation unless action is quickly taken to revive them.

Those of the Sami population who are ill or old end up in institutions where they, as Sami speakers, have little contact with the staff. This is due to the lack of social wellfare and health care services in the Sami language and of the near total lack of Sami speaking staff in the municipalities of the Sami homelands.The shortcomings evident in municipal social welfare and health care services could be remedied by adopting Sami cultural autonomy as the basis for the legislation, administration, provision, and monitoring of social welfare and health care services for the Sami.

Land Rights

In item 11 the Committee on the Elimination of Racial Discrimination expressed its concern over the shortcomings in the guarantees for the land rights of the Sami people and over the national and international economic activities (mining, logging, etc.) which threaten the Sami way of life and in item 23 proposed that the Finnish Government draft a clear policy on Sami land rights and on the ratification of ILO Convention No. 169.

In its present legislation Finland does not guarantee the Sami rights on land, water and natural resources; nor does it guarantee the Sami people the right to the livelihoods that are part of their culture. This applies both to the rights relating to the property of the Sami population and to their rights under international treaties.

As early as 1978 the Constitutional Law Committee of Parliament gave an opinion (PeVL 7/1978 vp) that it would promote the aims of the Convention on the Elimination of Racial Discrimination and increase the protection of law afforded to the Sami people if the law on the division of water areas, as far as the legal rights of Sami communities are concerned, were the same as those of other title holders. To date, this has not been implemented as far as the Sami water areas or fishing rights are concerned; however, the fishing rights are protected under the Constitution Act of Finland, as was noted by the Constitutional Law Committee of Parliament as early as 1978.

The situation is similar as regards the ancient Sami right of ownership to land - and the related rights to practice reindeer herding, fishing and hunting - which remains unsettled in law, even though the Constitutional Law Committee of Parliament has repeatedly urged the Government to present proposal to solve the question without delay. Yet another area where these rights remain to be settled is administration where the Sami do not yet have a special position.

In 1993 the earlier Sami Parliament was assigned the task of drafting a proposal on land rights, and the Parliament set up a working group for that purpose. The current Sami Parliament pursues the task, with a view to producing a proposal in the form of a Government Bill, which would in its part remove obstacles to the ratification of ILO Convention No. 169. In accordance with the views adopted by the Ministry of Justice and the Ministry of Labour' the Government is obliged to provide the Sami Parliament material support for its task. The Sami Parliament considers the concern expressed by the Committee over the Sami land rights justified. The question is before the Sami Parliament for settlement, but to date the Government has not allocated funds for the work. While the Government is yet to adopt a policy on Sami land rights, it has in any case entrusted the Sami Parliament with the task of drafting a proposal to solve the matter.

THE FUNDAMENTAL RIGHTS AND CULTURAL AUTONOMY OF THE SAMI PEOPLE

The legislation relating to the Sami population has recently undergone a profound reform. On 17 July 1995, in the context of the Fundamental Rights Reform (9fi9 95), a new Section 14 was incorporated in the Constitution Act of Finland which iii subsection 3 guarantees the Sami people as an indigenous people, the right to maintain and develop their own language and culture.

The reform came into force on I August, 1995. At the same time, a new section 51 u (t)73/95) was added to the Constitution Act which gives the Sami people, as an indigenous people, cultural autonomy in respect of their language and culture within the Sami homelands More specific provisions on the mattes were incorporated in the Act on the Sami Parliament (974/95), which applies to a body of representatives elected by the Sami people from among themselves and the powers of that body.

The laws on cultural autonomy came into force at the beginning of 1996. They do not apply to land rights or to the right to a livelihood.

Together with the Sami Parliament Act, the amendments to the Constitution Act considerably improve the possibilities of Sami people to develop their own language and their own culture on the basis of cultural autonomy. In practice, however, the implementation of the reform is being slowed down by a lack of resources and vocal anti-Sami activities.

The purpose of the Act on the Sami Parliament was to slightly widen the definition of Sami which is based on the Sami language. Because of a technical error which through oversight was left in the bill, the definition of Sami became so ambiguous that most non-Sami people in the Sami homelands have reason to believe that if they so wish they can seek to be defined as Sami. They believe that this gives them a right to decide on cultural autonomy for the Sami and to enjoy any other rights and benefits afforded to the Sami people.

The Sami Parliament, adopted on 1 March, 1996, a statement to draw attention to the shortcomings in the Act on the Sami Parliament. The Parliament rejects the new, wider definition of a Sami and demands that the definition based on language be restored to what was the definition in the Decree on the Sami Parliament (988/90) and still is in the Sami Language Act (516/91). If that definition is not restored, there is a risk that the wider definition in practice leads to the forced assimilation of the Sami with the majority population, against the purposes of the legislation on cultural autonomy.

In its statement the Parliament also requires that the laws relating to the electoral register for the Semi Parliament be harmonised with the safeguards under European Community law for the private life of an ethnic group.

RELATIONS BETWEEN POPULATION GROUPS

The unresolved nature of the Sami land rights and of rights to livelihood together with the problems relating to the Sami right to cultural autonomy and an ambiguous definition of Sami is about to turn the concept of cultural autonomy for the Sami people against the Sami. Some members of the majority population in the Sami homelands have organised into a registered group called *'Lappalaiskulttuuri - ja perinncdyhdistys r.y.'* (Association for Lapp Culture and Traditions).

The association works systematically against the Sami population and against cultural autonomy for the Sami people, attempting to prove its own members as the real Sami and the Sami people as self-seekers and fakes. The purpose of this activity is to undermine the identity of the Sami minority and to obscure and deny the existence of the Sami culture. The anti-Sami group is not interested in the language and culture of the Sami people. Instead, they try, against the wishes of the Sami population, to be elected to the Sami Parliament as Sami, in order to enjoy the economic benefits they expect to gain and to nullify the Sami cultural autonomy.

The means used by the group include threats made in public on civil war and violence. The group has publicly spread lies and defamatory comments about Sami people, their representatives, and officials in Sami administration, in addition to spreading anti-Sami propaganda.

The group has obstructed the work of the Sami Parliament by making requests to the police and the highest authorities exercising legality for investigations. and by submitting demands to the supreme courts for the cancellation of the results of the elections for the Sami Parliament and for the publication of electoral registers on the Sami, which records on ethnic origin.

The municipalities in the Sami homelands - Enontekio and Inari - where the Sami are a minority, have acted in support of the anti-Sami activities of the majority population. A number of news media and journalists, such as the largest provincial daily in Lapland, the major national daily newspaper and a national news agency, have since the spring of 1995 systematically transmitted anti-Sami propaganda in the name of the 'Lappalaiskulttuuri-ja perinneyhdistys r.y'. This type of activity seems to be increasing and taking harsher forms.

The Penal Law

To date, public prosecutors have not *sua sponte* held it necessary to, under the Finnish Penal Code, prosecute activities directed against the Sami. In addition, the Finnish Penal Code does not in every respect fulfill the obligations under Article 4 of the Convention on the Elimination of Racial Discrimination. The Article requires that the State Party declare illegal, among other things, the following acts:

- all dissemination of ideas based on racial superiority or hatred;
- incitement to racial discrimination against ethnic groups or individuals;
- all acts of violence and incitement to such acts against ethnic groups or individuals;
- assistance to racist activities, including financing;
- organizations which promote racial discrimination:
- organized and all other propaganda activities which promote racial discrimination; and
- participation in such organizations or activities.

All the obligations under Article 4 of the Convention should be incorporated in the Finnish Penal Law in the manner required by the Convention.

RETHINKING HUMAN RIGHTS FOR THE 21ST CENTURY

The proposed seminar seeks to rethink and regenerate fundamental concepts in the field of international human rights. Since the end of the Cold War, the promise of international human rights has captured the imagination of a global community of citizens, policymakers, scholars, and activists. Human rights law has become arguably the central normative framework for how states should treat citizens in an increasingly globalized world.

Human rights thus constitute an inescapable dimension of global political discourse, ranging from international diplomacy to economic development programs to grassroots community organizing.

However, despite its centrality, the dominant conceptual framework for human rights has important shortcomings. The proposed seminar examines those limitations and explores an alternative human rights framework based around the concepts of vulnerability and resilience.

The discourse of international human rights that was formed after World War II and institutionalized in major international treaties and declarations ascribes a universal set of rights to all humans no matter their sex, ethnicity, nationality, religion, or location. Central to this classic human rights discourse is a framing of circumstances in terms of a dialectic of rights and violations. The dialectic typically posits a perpetrator (usually a state) violating a victim's rights, which are endowed by virtue of the victim's humanity.

Human rights are generally understood and explained with reference to this dialectic. The rights/violations framework has many important qualities, including establishing concrete terms for identifying and criminalizing the abuse of power in specific domestic and international settings. The framework is also fundamental for "naming and shaming" practices that non-state actors use to bring human rights abuses to public attention.

However, the central framework for conceptualizing international human rights has critical limitations, which deeply affect scholarship, advocacy, and policy. First, the rights/violations dialectic tends to be a static, ahistorical, and acontextual paradigm for conceptualizing complex situations of suffering, repression, and violence.

Classic human rights discourse often produces reductionist, one-dimensional understandings of the forces that shape why human rights abuses occur. By extension, classic human rights discourse tends to strip away the conditions that precede, facilitate, and follow specific moments of abuse. In effect, then, despite being a major political, legal, and social paradigm in contemporary global affairs, classic human rights discourse is often too narrow an analytic tool for understanding complex problems.

Second, the dialectic of rights and violations tends to silence human agency. Classic human rights discourse often produces an image of pure victimization that sparks moral outrage, but often does so by relying on an image of a legitimate innocent victim helped by heroic outsiders. Such framing pushes to the margin the capacity for creative human agency and adaptation amidst regimes of abuse. It constructs victims as if victims were victims only, rather than resourceful people with responses and claims of their own. The framing also pushes to the margin the political thinking and claims of victims.

Third, the rights/violations framework tends to compartmentalize rights and to privilege civil/political rights over economic/social rights. Major international human rights instruments isolate specific types of violation. In so doing, they undermine connections between different forms of abuse and are in competition with a holistic, indivisible view of human rights.

The rights/violations framework also highlights more visible forms of violation that have a clear and direct perpetrator and that are associated with civil and political rights, such as the right to freedom of association and the right to be free from torture.

However, the framework sits less comfortably with social and economic rights concerns, such as access to water, food, health, and land, which are more invisible, constant, and structural forms of deprivation.

The proposed Sawyer Seminar will critically engage these and other concerns with what we label "classic" human rights discourse and practice. In addition, the proposed seminar will consider, discuss, and critically evaluate an alternative framework for human rights based on the concepts of vulnerability and resilience.

A vulnerability/resilience framework has several potential advantages over a rights/violations one.

First, the concept of "vulnerability" creates the conceptual and theoretical space for a historically continuous approach to human rights.

Vulnerability here means structural positions of weakness that foster or enable domination or transgression by others. Approaching human rights in this way focuses on broader contexts that produce—and outlast—moments of acute abuse. The approach can also be instructive for understanding how, when, and against whom human rights abuses occur.

Vulnerability also has the potential to establish a robust conceptual bridge between civil/political rights and economic/social rights. The divide between the two kinds of rights emerged during the Cold War and has been a central concern ever since. In most cases, human rights problems associated with social and economic rights are more conditions of deprivation and less specific moments of violation—the latter being typical of civil/political rights. A lens of vulnerability brings conditions of deprivation into focus, while also providing analytic purchase into when and how specific civil/political violations can occur.

The concept of vulnerability thus has the potential to support a holistic vision of human rights.

The Seminar will also examine whether a vulnerability framework provides greater insight into human rights problems of the future, in particular those associated with the environment, food, and health.

In addition to vulnerability, the Seminar will explore the concept of "resilience" as providing a richer, more historically perceptive understanding of the experience of human rights abuse than does "violation."

In most human rights situations, victims are resilient and creative; they do not require external actors to survive. Resilience is also central to the lived experience and memory of those who survive human rights abuse. Resilience thus highlights human agency as an integral part of human rights, and the framework challenges the "rescue paradigm" that requires external, often Western, intervention to end abuse.

In sum, the Seminar seeks to understand human rights as more than a static, sometimes reductionist, dialectic of rights and violations by introducing a more relational, historically situated, and dynamic framework. The point is not to reject the existing dominant human rights framework and its attendant institutions. Rather, the intention is to rethink fundamental concepts to reinvigorate our capacity to understand human rights. The Seminar will in turn explore whether a framework of vulnerability and resilience allows analysts and practitioners to contextualize and historicize abuse; to gain analytical specificity on how, when, and against whom abuse occurs; to recognize the importance of human agency in surviving human rights abuse; to create a conceptual foundation for linking civil/political and social/economic rights; and to set the stage for thinking about how human rights abuse will occur in the future.

Major Questions

The fundamental questions that will drive the seminar are the following. First, what are the limitations of classic human rights discourse and practice? In what ways does the dominant framework occlude certain forms of suffering and obscure certain forms of redress? In what ways does the rights/violations dialectic highlight certain human rights histories and downplay others? In what ways does the rights/violations dialectic privilege certain types of rights over others? Does the dialectic lead to certain types of representation of victims and perpetrators? At the same time, participants will be sensitive to the advantages of the current framework.

Second, what are the advantages and disadvantages of conceptualizing human rights in terms of vulnerability and resilience? What does a vulnerability and resilience framework contribute to our critique of (and appreciation for) a rights/violations framework? Does a vulnerability/resilience framework allow for a more contextual and historical approach to human rights? Does a resilience framework capture aspects of agency and survival that are real to many who suffer abuse? Are there concrete examples in which a vulnerability and resilience framework provides greater analytic insight than a rights/violations framework? Are there alternative frameworks for examining human rights that are different and more insightful from the two central ones discussed here?

Third, how does the alternative conception of human rights proposed here change the history of human rights? Is the classic human rights discourse a product of a particular historical moment, in particular the one that emerged after the end of World War II and, later, the end of the Cold War? If so, how would an alternative conception suggest a different historical trajectory of human rights?

Fourth, does a vulnerability and resilience framework anticipate future human rights challenges? For example, does the framework capture how humans can suffer in an increasingly globalized world? Will climate change present a new human rights challenge and will it require rethinking "violation" and "perpetrator," given the ways in which cause and effect are spaced geographically and across time? Can a vulnerability and resilience framework grasp the ways in which certain populations will suffer and survive climate change? If food security is a challenge of the future, how is food security related to other forms of structural vulnerability such as poverty? Can a vulnerability and resilience framework help us to understand who will face food security problems and who will not? The same question applies to questions of the right to health and the ability to access clean water, issues that are likely to be critical future human rights challenges.

Fifth, what does it mean to advocate for human rights in the 21st century? How would a vulnerability and resilience framework change the ways in which human rights practitioners approach specific cases? Would an alternative framework create different types of human rights reports?

Would an alternative framework allow practitioners to see patterns of human rights abuse in new ways? In what ways might a vulnerability framework enhance (or detract from) legal approaches to social and economic rights? In short, is the alternative conception a usable framework for practitioners or is the approach primarily in the service of scholarly approaches to understanding human rights?

Finally, in what ways does a vulnerability and resilience framework provide analytic purchase on age-old issues of the universality of human rights?

One of the central objections to the international applicability of human rights is that they are an imposition of priorities and values from the Global North onto the Global South. By the same token, the major drive for emphasizing social and economic rights has emerged from the Global South. Does the proposed alternative framework help reframe divisive debates about whose rights matter?

Comparative Case Research

The proposed Seminar will be divided into two distinct phases across an 18-month period. In the first phase, participants will examine the concepts, themes, and questions outlined above. In the second phase, participants will examine specific human rights topics and cases. The general model for each seminar will be to host three external speakers and two internal discussants from the University of Wisconsin, Madison campus. Each presenter will circulate a minimum of a five-page single-spaced memo prior to the seminar. All seminars will be open to the University of Wisconsin academic community as well as to the broader Madison community.

Institutional Resources and Suitability

The idea for the proposed Sawyer Seminar emerges from more than a year of reflection and programming among a network of scholars working under the umbrella of the Human Rights Initiative at the University of Wisconsin, Madison. Started in Fall 2007, the Initiative has sponsored to date a series of talks and workshops.

The Initiative's central objective is to enhance the intellectual community of diverse faculty who work on human rights and to establish curricular opportunities for undergraduate and graduate students. The core faculty on the Steering Committee is interdisciplinary. In addition to Scott Straus, the Committee includes Jo Ellen Fair (Journalism/International Studies); Susan Friedman (English/Institute for the Humanities); Francine Hirsch (History); Sharon Hutchinson (Anthropology/African Studies); Helen Kinsella (Political Science); Heinz Klug (Law/Center for Global Legal Studies); Steve Stern (History/Vice-Provost); and Aili Tripp (Gender and Women's Studies/Political Science). Florence Chenoweth, formerly of the Food and Agricultural Organization, is the Initiative's Managing Director and Lori diPrete Brown, Associate Director of the Center for Global Health, is active on the Committee.

The Initiative has a larger network of Faculty Affiliates, including scholars in Political Science, Education, Legal Studies, and the Romance languages. The Initiative also sponsors a Graduate Student Working Group, which meets regularly to discuss human rights scholarship and is planning a Spring

Symposium in 2009. These various groups of faculty and graduate students will constitute the principal constituencies for the proposed seminar. In addition, we expect that particular seminars will draw faculty and students with overlapping interests.

Projected Outcomes

In addition to fostering intellectual growth on the topic of human rights, the Seminar aims to produce one of two outcomes. The first outcome would be a collected volume based upon the contents of the seminar. The second outcome would be a series of monographs to be published as part of a new Human Rights Series at the University of Wisconsin Press. Gwen Walker, Acquisitions Editor at the Press, is currently in discussion with leaders of the Human Rights Initiative about creating the new series.

2

Climate Change and Environmental Refugees

The origin of environmental refugees can be partially blamed on our dependence on non-reusable energy which has caused increased gas emissions into the atmosphere. These energy bi-products include carbon dioxide (CO_2), methane (CH_4), nitrous oxide (N_2O), and fluorocarbons (HCFC, HFC, PFC). These gases, grouped under the term "Greenhouse Gas Emissions", have, since the 1990's, been identified as being responsible for climate change. CO_2 is the worst greenhouse gas with a 76% distribution in earth's atmosphere; the other gases follow with CH_4 at 13%, N_2O at 6% and HCFC, HFC, PFC running at 5%. Another reason cited in scientific literature to explain climate change is the earth's natural warming and cooling cycles. Deep-sea cores have revealed evidence that the earth warms and cools naturally on a 1,000 to 3,000 year cycle. The last cooling period may be the so-called Little Ice Age which began ca. 1100 CE. This cool period lasted for a few hundred years and then a warming trend again started; we may currently still be experiencing this trend.

The important question, however, is – whatever is causing the current climate change, be it greenhouse gas emissions and/or the earth's natural cycles — can we reverse this trend?

If we could stop all greenhouse gas emissions immediately, would it change anything? How long would it take to drop CO_2 back to the level of the early industrial period (circa 1900) when CO_2 levels sat at 290 ppm? Dr. Susan Solomon, an atmospheric scientist at NOAA's Earth System Research Laboratory and winner of the 2009 Volvo Environmental Prize, mentions in her pioneering study entitled Irreversible climate change due to carbon dioxide emissions that *"changes in surface temperature, rainfall, and sea level are largely irreversible for more than 1,000 years after carbon dioxide (CO_2) emissions are completely stopped"*. She also adds that *"increases in CO_2 that occur in this century* *"lock in" a sea level rise that would slowly follow in the next 1,000 years"*. If this is the case, humans must now be resourceful. The world's populations must decide upon a response to our changing climate.

Can we decrease the number of environmental refugees through Reforestation? The World Health Organization diagram below clearly indicates that climate change is a driving force that directly influences human populations world wide.

One of the major issues identified by this chart is desertification/land degradation. Forests are ecosystems that have sustained human life for eons. We use forests as a source of fuel for cooking and as a source for building materials. We also use forested lands as a source for many foods and medications. Wood, if managed properly, is very inexpensive, easily accessible and renewable but overuse of this natural resource has resulted in deforestation and the degradation of human environments. Reforestation is by far the most efficient way to tackle climate change and hence the most efficient way to decrease the number of environmental refugees. Prof. Norman Myers, environmentalist, authority on biodiversity and three time leading environmental prize winner, stresses in "Environmental Refugees: An Emergent Security Issue" that reforestation is the best way to embrace a sustainable development approach in countries facing difficulties in meeting human needs. Reforestation will "rehabilitate hydrological systems and watershed functions, and thus avoid floods and drying-out of river systems downstream".

Another proponent of reforestation is the 2004 Nobel Peace Prize Laureate Wangari Maathai's. She is now Deputy Minister for the Environment in Kenya. During her Nobel prize acceptance speech, she mentioned that *"A degraded environment leads to a scramble for scarce resources and may culminate in poverty and even conflict"*. She started a tree planting program in 1977 with just seven trees. This soon became the Green Belt Movement where Kenyan women are encouraged to care for seedlings obtained from native trees. These seedling are used as stock plantings for reforestation in villages. Numerous non-governmental organizations (NGOs) are now involved with reforestation in other developing countries, for example in South Korea, in China and in Turkey. The aims of reforestation projects are simple: restoration of water tables, rehabilitation of ecosystems, reduction of ambient air temperature and creation of buffer zones to decrease strong wind velocity. Reforestation with indigenous trees and shrubs also helps produce natural medicines, provide biofuel and other derived products such as honey.

One such successful reforestation program has occurred in Rio de Janeiro, Brazil. Started in 1986, it is called the "Community Reforestation Project"; the mandate of this project involved the correction of two important issues: slope stabilization and job creation. It focused on planting indigenous trees on steep hillsides inhabited by *favelas*. *Favelas* are spontaneous, unregulated settlements which consist of "*...an area, predominantly of housing, characterized by the occupation of land by low-income populations, precarious infrastructure and public services, narrow and irregular layout of access ways, irregular shaped and sized plots and unregistered constructions, breaking with legal standards*". To achieve their goals, the project concentrated on hillside improvements and on reforestation. They hired *favela* occupants to build infrastructures such as terraces, drainage canals and proper roads. These projects were specifically designed to stabilize the soil, increase water retention and reduce erosion on the steep hills. The local workers also acquired skills that were transferable to construction sites in lowland areas, providing future employment opportunities.

Instead of trying to organize massive reforestation, the project opted for planting small, scattered plots of land over a three to four year period. They also focused on the re-introduction native tree species well adapted to hilly terrain. The forest plan included planting fast growing leguminous trees mixed with native, exotic and fruit bearing trees. The project staff also began educational programs to raise the awareness of the favela populations, instructing them on how they could protect and maintain the new forested areas. Of 54 communities who participated in this reforestation project, only two recorded tree loss. This project resulted in a decrease of urban sprawl caused by the expansion of *favelas*, the regrowth of indigenous vegetation, an increase in available food, the return of native wildlife species, the regeneration of springs, the provision of shade and the improvement of air quality.

Can we decrease the number of Environmental Refugees through Adaptation? The term 'adaptation' when used in the area of climate change is the adoption of new behaviours that will help people cope with environmental changes. There are numerous smaller examples of adaption strategies, including selecting farm animals with limited water or foliage requirements, growing short season crops or drought and pest resistant crops, and changing irrigation techniques to reduce water requirements. There are also large scale adaption methods that could help us alleviate the affects of climate change including land reclamation and climate proofing.

ADAPTATION: LAND RECLAMATION

A large percentage of the global population (one out of 10 humans) lives in lowland areas with elevations at, or below, sea level. At the moment, 10% of the of the world's total population, and 13% of the world's urban population, live in lowlands; this represents circa 600 million individuals. With an increasing number of stronger storms, the disappearance of coastal forests and a general, slow rise in sea levels, erosion of coastal areas is inevitable. Unless internal policies, such as mitigation or migration to higher ground, are implemented, people living in lowlands will face constant serious threats to their livelihood. One adaptive approach to mitigate loss of coastal areas is land reclamation.

The main objective of land reclamation is to add more land for agriculture and human settlement. It is basically a survival strategy. Land reclamation is done by diking, creating embankments thick enough to prevent percolation of water at the base and high enough to prevent even the highest tides from washing over the top. This method must be implemented carefully and managed properly to prevent loss of inter-tidal zones such as wetlands and coastal forests (e.g. mangrove forests). These zones are responsible for natural evaporation processes that bring water from the ocean to the land and they play a large role in the absorption of much of the ocean's forces. Inter-tidal zones are also rich with food, important for humans and animals alike.

Diking can work as the natural landscape generally adapts to such changes, but if these changes are poorly designed or implemented too rapidly, they will adversely affect the ecosystem. Poorly researched, implemented and managed land reclamation may decrease the amount of rainfall, increase the frequency of dry weather, decrease biodiversity and deplete fish stocks. Red tides, which are caused by an accumulation of nutrients in river deltas, will also occur more frequently if there is a restriction in tidal movements.

Land reclamation is the adaption of choice in Singapore. Hazri Hassan, Singapore's Environment Minister mentioned that his country has opted to reclaim land and raise the elevation by 125 cm (four feet) above the highest sea level predicted by the UN over the next century. He feels that *"In Singapore, it is more cost effective to adapt than to retreat"*. Other countries which have also opted for land reclamation are: the Netherlands, who are already experts in this field, Japan, Hong Kong and Macau.

Adaption: Climate Proofing

Climate proofing is *"a decision-making process in land and water management, in which both risks and opportunities of climate change are taken into account. The concept of climate-proofing uses a combination of infrastructural and institutional measures in order to adapt to future climate change"*.

Global demography tells us the location of the most vulnerable geographical areas in the world. Two-thirds of the global population live in Mumbai, Shanghai, Karachi, Delhi, Istanbul, Sâo Paulo, Moscow, Seoul, Beijing and Mexico City. Five of these mega cities – Delhi, Istanbul, Mexico City, Mumbai and Sao Paulo — are located in vulnerable geographical areas already prone to flooding, storms and earthquakes. Other large cities such as Bangkok, Jakarta, Shanghai, and Hanoi are now facing the effects of climate change with extreme weather events.

Many cities are now beginning taking climate change seriously, creating strategies to increase their resilience to climate change. Milan, Tokyo, New York, China's Dong Tan, Hanoi, Singapore, and a few cities within the Metro Manila area have initiated climate proofing strategies that include monitoring and protecting water supplies, providing more efficient public transportation, reducing chemical and other wastes from industry, providing more green space and educating the public. These cities believe that they will decrease the effect of climate change by *"informing the population about the negative effect of climate change, providing more green space; using bicycles or walking more; and increasing the use of energy-efficient public transport vehicles"*.

Can we decrease the number of Environmental Refugees through Integrated Risk Reduction?

Integrated Risk Reduction is an adaptation strategy with the main goal of managing "current and future risks *"to ensure that these risks are reduced to an acceptable level. It is a holistic approach that is an integral part of sustainable development planing. This will have "long-lasting and environmentally sound, economically viable, and socially acceptable changes implemented at one or more of the following stages in the project cycle: planning, design, construction, operation, and decommissioning"*.

Change Adaptation through Integrated Risk Reduction, or CCAIRR, has a lot of potential especially in International Humanitarian Aid. Detailed information on this complicated and systemic adaptation strategy is available in a report entitled "Climate Proofing: A Risk-based Approach to Adaptation" which presents case studies done in the Cook Island and the Federated States of Micronesia (FSM). Readers will understand, how to apply risk assessment methods and tools, how to counteract some of the negative effect of climate change and how to request funding for climate proofing projects.

Relocation – the Final Option

Relocation is the *"transportation of people (as a family or colony) to a new settlement (as after an upheaval of some kind)"*. Whole scale relocation of a population would only be done when reforestation, land reclamation, climate proofing and Integrated Risk Reduction principles cannot be applied successfully because the land itself is threatened by ever increasing sea levels.

Since 1880, the sea levels have risen by 20 cm (8 in.) and it is predicted they will rise even more reaching 18cm to 59 cm (7in to 23in.) by 2100 in some areas.

Tuvalu, Vanuatu, the Marshall Islands, the Cook Islands, Fiji and the Solomon Islands, are all nations which are already affected by this rise. Two other island nations, Kiribati, which since 2000 had to evacuate two village islands, and the Carteret Islands, which had to relocate their population to Bougainville, were unable to mitigate, prepare, respond or recover from the effect of climate change. This is not all, the Maldives, in a matter of two weeks, just prior to 07 June 2009, lost 1,200 ft (366 meters) stretch of beach. Faced with this catastrophe, they decided to ask the international community for help. They would like either to buy land or islands to relocate their entire population or to be accepted as immigrants in a host country.

Tuvalu is an island nation consisting of nine atolls (representing an area of 10 square miles/26 square km) and a population of 11,800 people. A few years ago, Tuvalu began to experience stronger king tides. These are higher than usual tides that appear in February when the sun and the moon are aligned. King tides have a larger destructive potential than standard seasonal tides, often causing severe flooding. To date, main roads have been destroyed, croplands have been rendered unproductive, saltwater has intruded into groundwater and coconut trees are partially submerged. Faced with such destructive forces, Tuvalu decided to act quickly instead of waiting for international help programs.

The simply decided to leave their island nation altogether; they have asked the New Zealand government if they could emigrate. New Zealand is accepting limited numbers of Kiribati, Tuvalu and Tongan citizens every year and it may take six or more years for the whole population to relocate.

BEIJING LEANS ON NEPAL OVER TIBETAN REFUGEES

Under pressure from the Beijing, the Nepalese government has arrested hundreds of refugee Tibetans who cross the border from Tibet to escape persecution by the Chinese regime, refugees say, sending many back to China and earning objections from human rights groups. Kathmandu has also blocked human rights demonstrations led by Tibet support groups and has put a stop to all pro-Tibet activities, refugees who have made it to Dharamsala say, in particular on March 10 when Tibetans sought to commemorate the anniversary of the failed 2008 uprising in Tibet. The 20,000-odd Tibetans who live in Nepal are barred from holding any protests on its soil. Celebrations of the Dalai Lama's birthday are taken as anti-government activities by authorities. The tiny Himalayan nation is caught between Asia's two giants, India and China. Most Tibetans fleeing their homeland cross Nepal to India for shelter in the hill town of Dharamsala where their supreme leader and Tibetan Buddhism's highest religious figure, the Dalai Lama, took refuge in 1959 along with thousands of his followers with the consent of the Indian government. The pressure increased after China's army chief of staff, Gen. Chen Bingde, visited Kathmandu to pledge US$20 million in military and logistical support to the impoverished government of the tiny neighbouring country. It was not the first donation and is unlikely to be the last.

Chen led a 15-member army delegation to Nepal seeking to prevent any additional turmoil by the Tibetans in the country. It was the highest-level military visit from China to Nepal in more than a decade. The People's Liberation Army chief held high-level talks with Nepalese President Ram Baran Yadav, Prime Minister Jhalanath Khanal, Defence Minister Bishnu Paudel and his Nepalese counterpart General Chhatraman Singh Gurung, requesting that the Nepal government block movement of Tibetans across the border. "The purpose of my visit is to strengthen friendship and cooperation between Nepal and China," Chen told reporters. "This cooperation is not only conducive for our people but also for world peace and the Asia Pacific region." The Nepalese government reassured China of its support and commitment to the 'One China Policy,' and indicated it would bar anti-Chinese activity on its soil. It also agreed to "deepen military ties and ensure peace at the Tibet frontier." "General Chen expressed gladness at the steadily developing relation between two countries and said that economically developing and politically stable Nepal is important not only for the stability of Asia Pacific region but also for the whole world," according to a statement released by the Nepal's Prime Minister's Office in Kathmandu.

Over the years Beijing has vilified any labors by exiles to change the situation inside Tibet, which the Chinese occupied in 1950, terming the invasion a peaceful liberation of longstanding Chinese territory. Tibetans refugees started flocking to Nepal after the failed 1959 uprising against Chinese rule and the Dalai Lama fled over the Himalayas into exile in India. Interestingly Nepal has traditionally been more aligned with India. However, recent moves by Beijing, especially in infrastructure development, have made China a much more influential player in Nepal than India.

"We are treated as criminals in Nepal," said an exiled Tibetan who gave his name only as Tsering. "The Nepalese government does everything the way China wants it do, and we have even lost the right to do a peaceful march." He had lived in Nepal but fled to Dharamsala, he said. Nepal has launched many crackdowns since refugees staged protests against China in the run-up to the Beijing Olympic Games and the deadly 2008 Tibetan riots. Rights groups have declared Tibetan refugees in Nepal to be increasingly vulnerable and at risk of arrest and repatriation to China. Under pressure from China, Nepal has refused to recognize Tibetan refugees who arrived after 1989. They are not allowed to register marriages.

On any major anniversary of that Tibetan exiles attempt to celebrate, dozens are arrested, with reports that even monks are beaten, tear-gassed and kicked. The United Nations' human-rights office says some people are now being arrested purely "on the basis of their appearance" without being demonstrators. Recent Tibetan government-in-exile elections were treated as an unlawful activity by the Nepalese government, depriving a large number of people the right to exercise their vote. "We cannot allow such an illegal activity within our territory," Superintendent of Police Pushkar Karki, who also serves as chief of Kathmandu Metropolitan Police Circle, told local media. "We need to follow the government's policy." Last year the Wikileaks website disclosed US State Department cables from the embassy in New Delhi, alleging that China pays Nepalese police substantial money to arrest fleeing Tibetans.

"Chinese government rewards (Nepali forces) by providing financial incentives to officers who hand over Tibetans attempting to exit Tibet," said the cable, titled 'Update on Tibetan refugee flow.' Despite the restrictions, more than 2,500 Tibetans cross the border annually, embarrassing the Chinese government, which has sought to portray its administration in Lhasa as benevolent and dedicated to the welfare of the Tibetan people. The Nepalese government since the Wikileaks disclosures has come under fire for bending to pressure from the Chinese government in deporting exiles to China after previously having allowed them unhindered passage.

The Nepalese administration previously honoured a United Nations-brokered "gentlemen's agreement" between Nepal and the UN High Commissioner for Refugees to provide safe transit to Tibetan refugees who wish to travel through Nepal to Dharamsala, where thousands of Tibetan exiles have their base. That agreement has been abrogated under Chinese pressure. Watch groups such as Amnesty International, Human Rights Watch and the International Commission of Jurists have recently warned Kathmandu against breaching its international obligations, instead substituting "preventive arrests and policing restrictions on demonstrations and freedom of movement that deny the right to legitimate peaceful expression and assembly during anniversaries and festivals marked by the Tibetan community." "The link between China's aggression against Tibetans and Nepalese police actions has contributed to an environment of fear and insecurity in Nepal's Tibetan communities," the International Campaign for Tibet, the US based Tibet lobby said. Western countries, particularly the United States, have pressed the Nepalese government to soften its stance. The Undersecretary of State for Democracy and Global Affairs Maria Otero, the White House's special coordinator for Tibetan issues, visited refugee centres in Nepal in February where she met government officials to raise the issues of free passage and problems faced by Tibetans in Nepal itself. She pledged the continued support of the US government for the safety and welfare of refugees. "We have a longstanding policy, as you know, of supporting the needs of vulnerable refugee populations, and consider the Tibetan populations in Nepal to be particularly vulnerable. The U.S. continues to monitor the situation of both newly arriving refugees and the long-staying populations," a State Department official said in a statement on Feb. 18.

The Tibetan government-in-exile has appealed to Western nations to persuade the Nepalese government to allow the Tibetan activities and respect the human rights situation. Analysts believe China's long term plan is to neutralize Tibetans outside Tibet, leaving no voices to speak against them when the Dalai Lama retires completely. Rameshwor Acharya, the former Nepal ambassador to China, told reporters on March 24 that "China's concerns over Nepal are growing" and that "the visit shows that China wants the support of our army to control anti-Chinese activities following the resignation of the Dalai Lama." Thus many of the thousands of refugees wishing to leave their Himalayan homeland are finding that safe passage through Nepal is becoming increasingly unlikely and the future of Nepal's own Tibetan community appears uncertain as well.

TRANSIT OF TIBETAN REFUGEES THROUGH NEPAL

In recent years, between 2500 and more than 3000 Tibetans have runaway from Tibet via Nepal annually. In 1989, Nepal ceased to permit newly arriving Tibetans to remain or seek refuge in Nepal. By informal arrangement with the Office of the U.N. High Commissioner for Refugees (UNHCR), however, Nepal has acceded to an agreement whereby it facilitates the transit of new arrivals through Nepal, typically to Tibetan exile communities in India. This should operate as follows: Newly arriving Tibetans apprehended at the Sino-Nepalese border by Nepalese authorities will be denied entry to Nepal. Those apprehended within Nepal, however, will be turned over to the Nepalese Department of Immigration, which ordinarily then transfers them to UNHCR's custody. UNHCR cooperates with the Tibetan Welfare Office to make a preliminary determination that these Tibetans are "of concern to the High Commissioner," a broad, non-specific designation used by UNHCR to refer generally to asylum seekers, refugees, internally displaced persons, and others. UNHCR does not conduct refugee status determinations for Tibetans. Newly arriving Tibetans remain temporarily at the Kathmandu Tibetan Refugee Reception Centre, which provides them with food, shelter, and medical attention. By tacit agreement, they generally must depart for India within two weeks.

After UNHCR processes new arrivals and determines them to be "of concern," it ordinarily recommends to the Department of Immigration that they be issued an "exit permit," which solely provides Tibetans with the right to travel from the Reception Centre to the Indo-Nepalese border. It provides no right of re-entry, legal status, or protection of any kind. Indian border authorities typically take these permits, together with a small "fee," when Tibetan refugees cross into India.

Operation of the Gentleman's Agreement in Practice. The "terms" of the gentleman's agreement are rarely followed. In practice, most Tibetans arrive at the Reception Centre in Kathmandu independently, usually by one of two routes: via the Friendship Highway, which extends from Lhasa to Kathmandu, or via the Nangpa-la pass in the Solu Khumbu region of the Himalayas in northern Nepal. Tibetans often employ freelance guides to assist them in journeying safely to Kathmandu. Both the Friendship Highway and Nangpa-la routes are fraught with danger, including natural perils such as a lack of adequate food and shelter, particularly in the winter months, and the risk of apprehension by Chinese or Nepalese authorities. Contrary to the gentleman's agreement, Nepalese police increasingly appear to be forcing Tibetans apprehended within several days walking distance of the border to return to Tibet, often at great risk to their lives and safety. China treats these Tibetans as political dissidents. Reports of detention, torture, and other maltreatment by Chinese authorities are commonplace. Even without reference to the gentleman's agreement, the act of returning refugees to a state where their lives or freedom may be threatened violates the bedrock principle of international refugee law prohibiting *refoulement*.

The "Parties" to the Gentleman's Agreement. The gentleman's agreement requires the active cooperation of UNHCR, the Nepalese government, the Tibetan government-in-exile, and the United States Embassy in Kathmandu. UNHCR plays several vital roles. It provides a modicum of legitimacy that the Nepalese government deems crucial to its cooperation in implementing the agreement. UNHCR channels funds received from donor governments, particularly the United States, to support the day-to-day operation of the agreement. And until recently, it conducted missions to the border regions of Nepal in order to inform the Nepalese police of the terms of the agreement and international human rights law, and to monitor compliance. For reasons that remain somewhat unclear, the Nepalese government ceased to permit these missions as early as 1998, and

Tibet Justice Centre's research suggests that the consequences have included: an increase in instances of *refoulement* and other abuses, and an increasing lack of knowledge on the part of the police about their obligations under the agreement. The government of Nepal has acceded to the gentleman's agreement for two principal reasons: It requires UNHCR's assistance with its Bhutanese refugee crisis, and cooperation in the gentleman's agreement appears to be something of a *quid pro quo* for that assistance. It also remains under pressure from the United States, the European Union, and other foreign-aid donors to assist Tibetan refugees in need. Countervailing pressure from the Chinese government to repatriate Tibetans and tighten border control, however, has led to increasing incidents of non-compliance. The Tibetan government-in-exile, operating through the Tibetan Welfare Office in Kathmandu performs most of the practical work required to maintain the gentleman's agreement, including providing food, shelter, and medical care to new arrivals, assisting in their processing, cooperating with UNHCR to ensure that Tibetans reach the Reception Centre, and arranging for their departure to India. While the Nepalese government does not officially recognize the Tibetan Welfare Office because this would be perceived by China as an intolerable political statement, it relies on the assistance of the Tibetan Welfare Office to implement the gentleman's agreement. The United States Embassy in Kathmandu, finally, cooperates with UNHCR and the Nepalese government to ensure compliance with the gentleman's agreement. It participates in meetings with the Ministry of Home Affairs and works actively to advance the interest of the United States in ensuring the safe transit of Tibetan refugees through Nepal. The United States also provides much of the funding, channelled through UNHCR and the Tibetan Welfare Office, that supports the operation of the gentleman's agreement.

Allegations of Abuse and Refoulement by Nepalese Police. The gentleman's agreement cannot operate effectively unless the Nepalese police understand and comply with it. Tibet Justice Centre's research suggests that because UNHCR no longer carries out border missions, and because pressure from China appears to have caused the Nepalese government to cease to comply conscientiously with the gentleman's agreement, Tibetans arriving in Nepal increasingly face dangers of *refoulement* and, at times, police abuses. The police in Solu Khumbu reported receiving a directive from the Ministry of Home Affairs in Kathmandu instructing them to return apprehended Tibetans to the border, in violation of the fundamental principle of *non-refoulement*. In practice, this appears to have translated into a trend whereby the police will "encourage" Tibetans to return to China—at times, at gunpoint— and follow them back toward the border for several hours or even days. Because Nepalese police face dangers from the Maoist insurgency, and because, according to Tibet Justice Centre's interviews, they have not been receiving the stipends to which they are entitled for their cooperation, the police are reluctant to accompany Tibetans to the Kathmandu Reception Centre as required by the agreement. In some cases police reportedly resort to theft or extortion of money from new arrivals to supplement their salaries. In rarer instances there have been reported incidents of more severe abuses, including detentions, shootings, and rape. Resumption of UNHCR's border missions and greater efforts by the government to educate the police about the gentleman's agreement are vital steps to prevent such abuses in the future.

NEPAL'S SHIFTING MOVE TOWARD TO TIBETAN REFUGEES

From 1959 to 1986, but particularly in the existence before 1974, Nepal generally facilitated the work of foreign governments and humanitarian aid organizations involved in helping to resettle the first wave of Tibetan refugees. The Nepalese government viewed the settlements as an efficient long-term solution to a potentially serious refugee crisis. It therefore maintained a largely laissez-faire approach toward them. In 1975, UNHCR historian Louise Holborn lauded Nepal for "provid[ing] for the survival of thousands of Tibetan refugees" and giving them "the possibility of becoming contributing members of the societies of their host countr[y] while at the same time preserving their own identity."

Before 1986, the Nepalese government also tended to tolerate the arrival of new refugees for several reasons: It lacked the practical ability to enforce a strict border-control policy; fewer new refugees arrived during China's Cultural Revolution (approximately 1966-1979) because the Chinese government then maintained strict control over the movement of the Tibetan population; and in any event the Nepalese government did not perceive the number of newly arriving Tibetan refugees as a threat or a diplomatic liability during these years. The 1966 Agreement on Trade, Intercourse, and Related Questions Between the Tibet Autonomous Region (TAR) of China and Nepal permitted Tibetans living in the border regions of the

TAR and "religious pilgrims" to cross the Tibeto-Nepalese border without a passport or visa, provided they registered at the border. Nepal renewed this Agreement for another ten years in 1976. The Nepalese government's laissez-faire move toward Tibetan refugees began to change and tighten in 1986. In that year Nepal and China executed a new treaty that significantly restricted the ability of Tibetans to travel through or into Nepal. Soon after, in 1987 and 1989, the Chinese government suppressed a series of pro-Tibetan independence demonstrations in Lhasa. This led to a resurgence in the number of Tibetans fleeing into exile.

In 1989, pressure from the Chinese government and the growing number of new arrivals led Nepal to initiate a strict border-control policy. The Nepalese government made clear that it would henceforth refuse to accept or recognize new Tibetan refugees. At about the same time, however, UNHCR re-established an office in Kathmandu to assist Nepal with its Bhutanese refugee crisis. With the tacit approval of Nepal's Ministry of Home Affairs, UNHCR also began to facilitate the transit of newly arriving Tibetan refugees through Nepal to India. The funds for this informal and low-profile operation initially came from the United States, which provided an "infusion of $100,000 in 1991" to support the "refugee protection and processing program." The circumstances and legal status of Tibetans residing in Nepal, and the current operation of this informal "protection and processing program," are the twin subjects of this report.

TIBETAN REFUGEES' RIGHTS AND SERVICES IN INDIA

The Chinese assault of Tibet, and the Dalai Lama's following decision to flee in 1959, resulted in the mass influx of Tibetan refugees into India that continue today. It has become clear to the Indian government, as well as to the Tibetan community in exile, that repatriation is unlikely in the near prospect. Consequently, an evaluation of India's protection of, and assistance for Tibetan refugees is necessary to ensure their treatment is in accordance with international standards. Unfortunately, such an assessment shows that there is a lack of proper protections and services; this ought to be of particular concern to the international community as India has served on the Executive Committee of the United Nations High Commissioner for Refugees (UNHCR) since 1995.

How can a country be trusted to lead an international organization dedicated to the protection of refugees when it has such unstable and inadequate refugee policies itself? Central to the assessment is India's lack of legal compulsion to provide refugee protections or services. This is largely a consequence of India's abstention from the 1951 Convention relating to the Status of Refugees and the subsequent 1967 Protocol, which, in addition to defining a refugee according to international law, outlines certain rights and services host states must afford them. With the exceptions of the customary international practice of *non-refoulement* and the obligation to provide asylum as outlined in Article 14 of the Universal Declaration of Human Rights (UDHR), India does not have international, regional, or national obligations to refugee populations on its soil (Feller 2006). Despite this, and despite the nation's refusal to accede to the 1951 Convention, India argues its current refugee policy is in accordance with international norms. However, India has no national refugee law, and no regional agreement on the matter exists.

The only mention of foreigners in Indian law are in the Registration of Foreigners Act of 1939, the Foreigners Act of 1946, and the Foreigners Order of 1948— all of which employ the term *foreigner*, broadly defined as "a being who is not a citizen of India." Among other practices, the 1946 Act and the 1948 Order allow the government to limit employment opportunities, freedom of association, and a foreigner's possessions. These pieces of legislation also restrict movement inside India and prohibit select activities, most of which are political. In fact, Tibetan refugees are only allowed in India on the condition that they abstain from political protests against China – a clear suppression of their right to political expression as afforded by the UDHR. This political suppression was particularly evident during the 1991 visit of then Chinese Prime Minister Li Peng, during which Tibetan refugee leaders were arrested and placed under police surveillance to curtail protests. A similar occurrence was witnessed during Prime Minister Wen Jiabao's 2010 visit. Human rights law grants refugees fundamental civil and political rights, yet the Indian government denies these rights to Tibetan refugees.

India's lack of legal obligation allows it to change refugee policies at will, often to reflect its current political interests. The result is a rather unstable refugee policy, exemplified in the disparity of rights and services afforded to Tibetan refugees according to their date of arrival in the country. The Indian government regarded the first wave of Tibetans arriving in 1959 with the Dalai Lama as refugees and granted them legal asylum. They were allotted land and housing – a privilege not afforded to other foreigners and one that would later be denied to newly arriving Tibetans. More importantly, these refugees were automatically given a Registration Certificate (RC). According to the Tibetan Government-in-Exile, the Central Tibetan Administration (CTA), the RC is a legal document issued by the Indian establishment that allows Tibetan refugees "the right to enjoy all the privileges enjoyed by any Indian citizen except the right to vote and work in Indian government offices" (Immigration and Refugee Board of Canada 2009).

While many question the extent of the 'privileges enjoyed' by RC-holding Tibetans, the value of the RC is not contested. The RC allows Tibetans to legally travel and work within the country, serves as an identity document, and is a pre-requisite for the Identity Certificate, which is necessary for international travel. Tibetan refugees arriving between 1959 and the 1970s also benefited from services provided by the UNHCR, which despite its tumultuous relationship with India had a temporary presence in the country. The rights and services afforded to Tibetans inward after the 1970s are scarce and indicative of a changing Indian policy—arguably in an effort to preserve Sino-Indian relations.

In 1963, the Indian government ceased to legally recognize arriving Tibetans as refugees. Consequently, those arriving after 1979 (including some arriving in the late 1960s) have had greater difficulty acquiring RCs. The newer process to acquire an RC is arduous and particularly concerning because employment, residency rights, and international travel are contingent upon this document. In addition, while the Indian government provided land to the original refugee community, newer arrivals have not been as fortunate. The denial of land by the government is troublesome considering that farming is chief to a Tibetan's income and Indian law prohibits foreigners from purchasing land. Additionally, the UNHCR closed its office in New Delhi and discontinued its aid to Tibetan refugees in 1975. As a consequence of the Indian government's lack of assistance, the CTA (with substantial assistance from NGOs) has had to shoulder the burden of providing for these refugees. Although the CTA is able to cover basic needs, they are overwhelmed and are unable to effectively address the broader social welfare of this burgeoning community (Adams 2005). Even those with the RC carry on to struggle with their refugee status.

Unable to gain Indian citizenship, they cannot vote, work for local or federal governments, carry an Indian passport, or legally purchase land. Although Tibetans possessing an RC can legally work in pre-approved segments of the economy, many employers often discriminate against them in favor of Indian nationals. This resentment originates from the belief that Tibetan refugees are adversely affecting job opportunities and that they benefit from elevated standards of services not afforded to Indian nationals. While it is believed that communal relations between the Tibetans and the Indian host community have been improving, there have been isolated reports of aggression and discrimination that have largely gone uninvestigated by local police. To further exacerbate this issue, Tibetans are not represented in the local and central legislature and thus do not have readily available avenues for redress. In 1949, when the United Nations debated legal protection versus material assistance for refugees, India argued that "the guarantee of legal rights without concomitant material assistance was a hollow concept" (Oberoi 2006, 22).

However, it has become clear that as Tibetans continue to cross the border, India has not only decreased the rights afforded to these newcomers, but too the services provided to them. Given that repatriation for this refugee community does not appear possible in the immediate future, India needs to concern itself with improving its policy toward this population— thereby affording refugees the proper protection and treatment per international standards. Signing the 1951 Convention would be the best avenue by which to accomplish this as it will lessen the burden on India by affording it assistance and services from the UNHCR and HUMAN RIGHTS & HUMAN WELFARE member states. Above all, India's lack of legal obligation should not be understood as an absence of Tibetan refugees' entitlement to proper services and protections.

INDIA: SITUATION OF TIBETAN REFUGEES

The Central Tibetan Administration (CTA), the administrative branch of the Government of Tibet in Exile, states that 31,293 Tibetans arrived in India between 1979 and 1997 (Central Tibetan Administration Apr. 1997, 3; USCR 1997). According to Maura Moynihan, a consultant to Refugees International, who testified before the US Senate Foreign Relations Committee Hearing on Tibet in May 1997, approximately 25,000 Tibetans arrived in India between 1986 and 1996, increasing the Tibetan population in India by 18 per cent. In their 1999 annual report, the US Committee for Refugees (USCR) states that 3,100 new Tibetan refugees arrived in India in 1998 alone (1999). Previous USCR annual reports state that 2,000 and 2,843 Tibetans arrived in 1997 and 1996 respectively. The increase in the number of Tibetans in India since 1979 has overwhelmed the resources available within the Tibetan community and the Central Tibetan Administration.

According to the Government of Tibet in Exile, the most fresh Tibetan arrivals in India include political prisoners and prisoners of conscience, monks and nuns escaping religious persecution, pilgrims hoping to meet the Dalai Lama, people seeking to visit family members living in India and youth and young children in search of an education in Tibetan culture and language which is not available in Tibet. Approximately 45 per cent of arrivals in India since the 1980s are monks or nuns. Moynihan stated before the Senate Committee that 30 per cent are children seeking an education; according to Central Tibetan Administration statistics, 44 per cent of arrivals are between the ages of 14 and 25 and 17 per cent are 13 years old or less. According to USCR, the Central Tibetan Administration encourages Tibetans who come to India on pilgrimages or to study, to return to Tibet afterwards to maintain the Tibetan population in Tibet in the face of increasing Chinese migration and to ease the pressure on Tibetan settlements in India (1999). In January 1995, the Central Tibetan Administration established new regulations concerning the length of time recent arrivals in India could remain in the country. According to *The Tibet Journal*, monks between the ages of 16 and 25 can stay in India for six months before having to write and pass their college exams. Those who fail the exams are obliged, by the CTA, to return to Tibet.

Tibetans between the ages of 6 and 13 can stay and study in schools, 14 to 17 years-olds can study in the Tibetan Children's Villages, and those between the ages of 18 and 30 can study for one year, after which they are obliged to return to Tibet. Tibetans who can prove that they were subject to harassment in Tibet are exempted from the above rules.

Of the 2,843 Tibetans who arrived in 1996, the Central Tibetan authority requested 1,200 to go back to Tibet because the resources required to help them settle in India were non-existant (USCR 1997). Further information about those who were requested to return to Tibet was not available in the sources consulted by the Research Directorate at the time of publication.

In their 1997 report on India, USCR states that 60 per cent of Tibetans who arrived since 1979, have stayed for at least "some period" and an unspecified number returned after completing their studies. Tibetans who return from India to China often face harassment and work and travel restrictions (US Senate Foreign Relations Committee Hearing on Tibet 13 May 1997).

According to the Tibet Information Network (TIN), in March 1997, more than 60 Tibetan tour guides in Lhasa were prohibited from working in the travel industry after having illegally left Tibet for India without permission (27 Mar. 1997). In 1996, a returned musician was accused of being a spy and sentenced to 18 years in prison.

Further information on this case could not be found among the sources consulted by the Research Directorate. In 1993, Chinese authorities imprisoned a Tibetan tour guide who had returned from India. The guide was jailed for eight months for "stealing state secrets" and was only released after the international community pressured the Chinese government. In the same article, TIN stated that since 1995, increased security in Tibet has allowed police in Tibet to identify most returnees; many of whom are "detained and interrogated for up to three months upon their return to Tibet.

COUNTRY SITUATION IN INDIA

Housing

In the original agricultural settlements, the majority of which are located in southern India, each family of five was given five acres of land and a two to three-room house. Because of the natural growth of the Tibetan population and the influx of new Tibetans, a serious problem of overcrowding in the settlements has developed. An average of eight people are housed in each unit which was designed for five; in some cases there can be as many as ten. In many cases, the original houses have not been renovated since the 1960s even though they were designed and built as temporary structures. According to the senior researcher with the INS, as of 1996 no new houses had been built since 1959 in at least two settlements (22 Dec. 1999). Tibetans who arrived in India after the initial settlement did not receive official refugee status from the Indian government; they were neither allocated land for housing or farming nor provided with government assistance. Because of a lack of resources and the remoteness of some of the scattered communities, the approximately 10,000 unsettled Tibetans do not have adequate housing and often live in poor conditions.

Education

The Government of Tibet in Exile estimates that 80 per cent of Tibetan children attend Tibetan schools. The percentage of children being educated rises to an estimated 85 to 90 per cent of the school-aged population when the number of children attending non-Tibetan schools is included. There are three types of Tibetan schools in India: those run by the Central Tibetan Schools Administration (CTSA), an independent institution under the Ministry of Human Resource Development of the Government of India; those run by the Department of Education of the Central Tibetan Administration; and those run by private charitable organizations including the Tibetan Children's Village and the Tibetan Homes Foundation.

Like other facilities, the Tibetan education system has been stressed by the growth of the Tibetan community and the continual arrival of new Tibetans. In a 1996 interview with USCR, the Dalai Lama stated that Tibetan schools have been "seriously affected" by the influx of new school-aged Tibetans.

According to Maura Moynihan, the schools in Tibet are "seriously overcrowded" and are in need of textbooks, educational materials, and space for classes and dormitories in order to accommodate the increasing number of Tibetan children seeking an education.

Although a limited number of scholarships are available to Tibetan students who wish to further their education, the majority of Tibetans who graduate from high school are unable to continue studying due to financial constraints. Through the CTSA, the Indian government grants 20 scholarships a year to Tibetan students graduating from the CTSA schools while the Indian government reserves several seats in various Indian institutions for Tibetans studying engineering, medicine, pharmacy, and printing technology. The Indian government also provides two scholarships a year to Tibetans studying medicine or dentistry (INS 22 Dec. 1999). The Department of Education of the Central Tibetan Administration, the Tibet Children's Village and the Tibetan Homes Foundation also provide a limited number of scholarships for Tibetans to study either in India or abroad. The Tibetan community in India has also founded monastic Tibetan universities, including Deprung University, located in Doeguling, Karnataka (IPS 15 Jan. 1999).

In addition to the limited ability of Tibetan students to study at a post-secondary level, the Tibetan Settlements in turn, have a limited ability to absorb graduates from colleges and technical institutes. As a result, educated Tibetans often have to look for work outside of their own communities and settlements.

Health

The Department of Health of the Central Tibetan Administration is responsible for managing the Tibetan health care system in India and has established Primary Health Care Centres in almost every Tibetan community. According to the senior researcher with the INS, these 65 centres follow guidelines set by the World Health Organization (WHO) (22 Dec. 1999). In addition to the Primary Health Care Centres, the Central Tibetan Administration runs eight hospitals and 37 traditional "Tibetan medicine" clinics. The Tibetan Medical and Astrological Institute (TMAI) provides curative services, trains doctors in traditional Tibetan medicine, operates a pharmacy where Tibetan medicine is made and conducts research involving Tibetan medicine. Physicians working in the traditional clinics in the settlements were trained at TMAI. In 1993, the Department of Health of the CTA established a medical treatment and rehabilitation program for torture victims. The department also provides vaccinations, especially against tuberculosis, for new arrivals and food to newly-arrived women who are pregnant.

Many Tibetans in India suffer from gastro-enteric, diarrhoeal, skin, and respiratory diseases. These afflictions are mainly the result of poor sanitation and a lack of hygiene, which are exacerbated by the inadequate water supply available in both the settlements and the scattered communities. According to the Government of Tibet in Exile, the settlements are, on average, able to fulfil two-thirds of their needs for water for drinking and washing.

Some settlements are only able to meet half, or less than half, of their drinking water needs. Tuberculosis (TB) is a serious problem in the Tibetan communities of India. Although 35,000 cases of TB have been reported in the Tibetan community since 1959, the Government of Tibet in Exile states that the detection and treatment of TB has improved in recent years. Overcrowded housing conditions exacerbate the spread of TB and other health related problems. However, the health of Tibetans in the Hunsur settlement has improved with lower incidences of ailments caused by poor sanitation.

Traditional Learning Centres

Monks and nuns constitute about 45 per cent of all Tibetans who arrived in India since the 1980s. Monastic students constituted 60 per cent of all Tibetan arrivals between 1986 and 1996 (INS 22 Dec. 1999). Although some Tibetans return to Tibet after completing their monastic studies, many remain in India (USCR 1997) and consequently the monastic population in India has more than doubled since 1980. This increase has strained resources in monasteries and nunneries and has led to overcrowding and an increase in incidents of disease. Because monks and nuns do not participate in income-generating activities such as farming or carpet-making, they are dependent on the rest of the Tibetan community, international organizations and NGOs for support.

Employment

Agricultural production and animal husbandry explanation for 30 per cent of primary employment among Tibetans in India. This proportion rises to 50 per cent of economic activity in the settlements. Trading and sweater-selling account for another 30 per cent of activity in the Tibetan community as a whole and for 40 per cent of economic activity in the scattered communities. The rest of the Tibetan population relies on the service industry, handicraft making, carpet weaving and other cottage industries and employment in the Central Tibetan Administration for income. As many of the settlements are located on drought-prone land and only five per cent of the total amount of land in the settlements is irrigated, there is only one crop per year. Therefore, many Tibetans who work in agriculture also spend several months a year selling sweaters in cities outside the settlements to supplement their income.

Tibetans are free to work in the Indian economy, however, as non-citizens it is often difficult for them to find jobs (USCR 1997). Thubten Samdup, president of the Canada-Tibet Committee, states that access to employment is limited since employers often prefer to hire Indian nationals (21 Oct. 1999). Recent arrivals have greater difficulty in securing employment due to the fact that they do not have officially recognized refugee status (INS 22 Dec. 1999). According to the Tibetan Welfare Office of the Department of Home, which is part of the CTA, Tibetans do not have the right to hold jobs in the Indian government (1998). Although many Tibetans are economically self-sufficient (USCR 1999), the unemployment rate for Tibetans between the ages of 16 and 50 is 18.5 per cent. According to USCR, elderly people, woman-headed households, and recent arrivals continue to struggle economically (1999). In her testimony, Maura Moynihan quoted a survey conducted by Indian sociologists which estimated the average annual income of Tibetans to be US$150, as compared to US$350, for an Indian national.

Relations with the Local People

In an interview with USCR, the Dalai Lama stated that in general, relations between Tibetans and Indians are good(Oct. 1996). Sources state that the local population does not resent the Tibetan community, although there have been isolated incidents of anti-Tibetan violence. In November 1999, it was reported that the Dalai Lama was bearing in mind relocating some offices of the Tibetan administration and his personal residence to the Faridabad region "as the growing tension between the locals and the Tibetans is becoming a cause of worry " (Tribune 25 Nov. 1999). However, one week later, it was reported that the Dalai Lama had "ruled out" moving to Faridabad. The tensions which prompted consideration of the move occurred in July 1999, when approximately 140 Tibetan shops and market stalls in Manali were attacked and burned by Indian rioters after an Indian youth was killed by a Tibetan, following a disagreement between the Tibetan and three Indians. According to a report in The Tribune, a well-known Tibetan market was "razed to the earth" during the riots. The deputy police commissioner has requested a magisterial inquiry into the rioting and looting and has stated that efforts will be made to provide protection to the Tibetans. The Tribune described the situation in Manali as "tense, but under control". In a similar incident in 1994, an Indian mob burned Tibetan offices in Dharamsala after another Indian youth was allegedly stabbed and killed by a Tibetan.

In 1995, an anti-outsider campaign and a motion by the state of Arunachal Pradesh to expel approximately 12,000 Tibetans carried a lot of popular support, although the federal government was angered by the local government's decision. Both the US Department of State and the Dalai Lama stated that despite random and isolated incidents, Tibetans in India are able to lead peaceful lives. Thubten Samdup, of the Canada-Tibet Committee, emphasizes the point that India is a tolerant and compassionate country that has gone out of its way to accommodate Tibetans (21 Oct. 1999).

IMMIGRATION OF TIBETAN REFUGEES IN INDIA

In the initial stage, to accommodate and help Tibetan refugees, the government of India with assistance of the state government of Assam and West Bengal set up various transit camps. In the camps they were given all the necessities. Despite of assistance, Many Tibetan refugees died as a result of living in these highly crowded and unsanitary conditions. This unfortunate state of affairs provoked urgent needs of a long-term rehabilitation program that would serve to bring all the refugees into homogenous Tibetan communities that will facilitate them to perpetrate their culture and traditions and thus preserve their national identity. Thus His Holiness the Dalai Lama requested the Government of India to resettle the Tibetans in agricultural-based settlements as most of Tibetans were familiar with agricultural activities. Pt Nehru the then Prime Minister of India accepted and wrote to the government of Indian states, requesting them to make land available to refugees. Accordingly Government of Mysore (as Karnataka state was called at that time) allotted nearly 3,000 acres of land at Bylakuppe in Mysore district in Karnataka in 1960 and the first ever Tibetan exile settlement namely Lugsung Samdupling came into existence in 1961. A few years later another settlement namely Tibetan Dickey Larsoe also called as TDL was established. After that more and more Indian states provided land for Tibetan refugees. The government of India with requests from His Holiness theDalai Lama builds special schools for Tibetans that provide free education, health care and scholarships for those students who excel in schools. At present there are 35 different Tibetan refugee settlements in India. These settlements are further classified into Agricultural base, Agro-industrial Handicraft based and scattered communities. The total refugee population figures in India according to the report of the survey conducted by the Office of His Holiness theDalai Lama and Department of Home put it at 104686. The Indian government initially provided the Tibetan refugees with excessive amenities including legal Residential Certificates and Identity Certificates.

The second phase of Tibetan refugee mass departure to India was started in late eighties and continues in steady stead till late nineties. Indian government admitted arrival of 25,000 Tibetans refugees between the years of 1986 and 1996 from Tibet. With amplifying power and dominance of Communist China in the stage of world politics and the Indian government, in an attempt to improve its relations with China, declined assistance to these Tibetan refugees, and refused to grant them new allotments of land, denied these new arrival Tibetans both residential and identification certificates. The statement made by Exile Tibetan Government in late eighties and early nineties says "Tibetans who arrived in India after the initial settlement did not receive official refugee status from the Indian government; they were neither allocated land for housing or farming nor provided with government assistance." Additional to this, a analysis through various figures issued by Tibetan Refugee Reception Centre, a branch of Security department of Exile Tibetan Government who deals with new arrivals from Tibet, states that in an average 2500-3000 Tibetans come to India. Out of these, average of more than 60% stay back in India. This overflowing of Tibetan refugee in India and changes in the treatment of Tibetan refugee in India by Government of India, led to various problems. Central Tibetan Administration of His Holiness the Dalai Lama (CTA) argues that this has lead to overpopulation, unemployment, and food shortages in the Tibetan refugee community. Among them the unemployment among the large youth population seems to be a growing challenge that Tibetan refugee community in India face, as the possible impact it may have on the future of Tibetan fate in general and present refugee community in particular seems highly disturbing.

AIDS IN INDIA AND ITS POTENTIAL IMPACT

Since HIV takes from 5 to 8 years to obvious as full-blown AIDS, the final repercussions of the epidemic have yet to be realized in either the Indian or in the Tibetan refugee community. This can be contrasted, for example, with the situation in Africa, North and South America, and Europe, where the impact of AIDS was experienced in the mid-1980s. According to many health workers, social awareness of the reality of AIDS in India is a relatively recent phenomenon: "Sunlit Solomon, manager of the YRG Centre for AIDS Research and Education, recently found that among truckers in Madras who requested testing because they have Sexually Transmitted Diseases (STDs) and are concerned about their high-risk behavior, HIV seroprevalence rates rose from nearly 60 percent in 1993 to 91 percent in 1995. Also at risk are those who are not part of the high risk groups — people who need blood transfusions."

A Bombay study found that "86 percent of the commercial blood donors screened in 1992 were seropositive. Some commercial blood donors engage in unprotected sex to earn income; others may have been infected through unsterile plasma-phersis equipment (blood transfusion equipment)."

The transmission of HIV in India is also happening in a country that has very high rates of other potentially-terminal infectious diseases. Tuberculosis, leprosy, malaria, hepatitis, hemorrhagic fever and encephalitis are common in India and command greater attention due to their long history in the sub-continent. There are some "20 lakh (200,000) deaths a year from TB alone." It is also estimated that "46% of those who are HIV positive and are in the South and Southeast Asia region also have TB." Unfortunately, HIV serves as a magnet for TB since the immune system is impaired once a patient is infected with HIV. This is causing additional problems in controlling the spread of TB in India. Furthermore, many in India do not even realize that they have AIDS because their symptoms are so similar to those of other infectious diseases.

On the positive side, the collision of AIDS is producing a kind of revolution in India. Open discussion about sexuality, which was previously considered taboo, is now a socially accepted form of discourse. Evidence of this is found in (Indian) English-language women's magazines. Indeed, AIDS is forcing women to re-examine their role in traditional Indian culture. To be an 'ideal wife' endowed with the attribute of unquestioning subservience is now a dangerous role to assume. Another example of change is found in Indian cinema: two films in the 1996 Montreal Film Festival were about subjects that were traditionally considered off-limits: lesbianism and transsexuality. But the long-ingrained traditional customs and mores will continue to exact a significant toll upon Indian society before they are abolished or altered through cultural reformation. India is an ancient and multidimensional culture. Cultural values proscribe sexual relations. Given the close connection between religion and culture, it should come as no surprise that HIV is challenging traditional sexual customs. Though India has numerous tribal communities which also have their own specific practices, the following are some of the most common religio-cultural sexual practices in India:

The tradition of initiating poor, low-caste girls into the devadasi sect. Dedicated to the goddess Yellamma when they are young, devadasis historically served priests, assisted in religious ceremonies, and cleaned temples. With the decline of temples as power structures, however, their role was stripped of much of its religious import and they found themselves selling sex to survive."

Another cultural sexual phenomenon is that of the Hijras. "Hijras are eunuchs and transsexuals who sell sex for a living." Surveys done by Ram

Pal Vasisht, director of New Dehli's AIDS Control Cell, suggest however that "one in three hijras there (Chitti-Pournami festival) is infected with HIV." Dangas: "Dangas are men who 'become' woman for the duration of the (Chittirai Pournami festival) festival... they will don saris, elaborate wigs, and bright plastic jewelry. They also flirt, saunter, and sell sex to other men attending the festival."

Matas: "A beautiful `harijan', or a low-caste child, who attracts the attention of the village landlords and is chosen to become a mata or `sacred mother'. She is then dedicated to a life meeting her patrons' sexual needs: she is bathed ceremoniously in the temple pond, paraded through the village, and then sent to a seasoned mata who tutors the girl in her new role." These practices will need to be reformed in light of the reality of AIDS; but, as evidenced in the rest of the world, it will take time, and much education, to change sexual behavior which has been considered normative for so long.

TIBETAN REFUGEES IN NEPAL

During his fresh visit to Nepal, the South Asia Forum for Human Rights (SAFHR) took a bold initiative and arranged a visit to Pokhara, situated at 200 kilometre west of Kathmandu. Pokhara means "land of lakes". It is a beautiful hilly area situated at an altitude of 827 metres from the sea and is about eight hours drive from Kathmandu. Pokhara records the highest rainfall in Nepal and has several beautiful lakes offering an eye-catching view of Himalayan peaks.

The journey through the spiral roads running through the very high hills with evidence of land-slides towards Pokhara was very exciting. There was a wonderful reception from the refugees. Twenty-nine participants from eight countries working for human rights were in the list to visit the old Tashi Palkhiel Tibetan Refugee camp.

Most of the Tibetan refugees were lucky enough to have escaped across the borders into Nepal to live in settlements in Kathmandu' Budhanath district and the remainder in Pokhara, Baglung, Mustang, Taplejung, Manang, Rasuwa, Solukhumbu and Lalitpur. Virtually, Nepal is now the home of 18,000 formally registered Tibetan refugees (unofficially more than 20,000), who are known to have fled to Nepal in 1959, the year when Tibet's spiritual leader Dalai Lama fled to Dharmasala, India. Most of them are actually children and grandchildren of the refugees. In 1960 the Nepalese government arranged to provide Tibetan refugees with land. They established four temporary settlements. Dhialsa in Solu Khumba Mountain, Dhorpatan in western Nepal, Jawalakhel in Kathmandu and Tashi Palkhiel in Pokhara. As the Tibetans have no right to buy property, Nepal Red Cross (NRC) purchased the land with the donations from UNHCR.

There are four Tibetan refugee camps in Pokhara and the one visited by us, Tashi Palkhiel Tibetan Refugee Camp was established in 1962 in cooperation with the Swiss development Corporation and the government of Nepal. There are about 1000 refugees in the camp with facilities of a school, old age home and a beautiful monastery. The camp in Pokhara looks like a small Tibet. The refugee houses were minute, too closely constructed by less stable materials. Food and water supply was adequate but the source was not safe for drinking. Medical care was given by only a small dispensary run mainly by a female nurse of Tibetan origin, still the Tibetan preferred not to visit the dispensary but to depend on their traditional treatment. Dysentery, diarrhoea, skin infection was not common. There is no formal training for the women who are mainly engaged in the domestic activities, child care and sales women.

Refugees are not allowed to dabble in political activities, or do business outside of their area, though some are working outside, mainly in Kathmandu having valid ID card issued before 1989. The rest who came after that time cannot assimilate and find secure employment without formal legal standing in Nepal. Tibetan refugees have no right to own property or business, cannot own houses, vehicles, land and other property in Nepal. Legal and social discrimination severely curtail their ability to secure employment. They are not permitted to acquire residential and other legal documents. The displaced Tibetan nationals have experience harassment, extortion and deportation when they are without valid evidence. Tibetan refugees cannot travel to certain restricted regions of Nepal, and it is difficult to get travel documents for crossing international borders.

In recent years, refugees have come at a rate averaging eight person per day. The new arrivals are fed and housed at a transit camp in Kathmandu run by the administration of Dalai Lama. After their arrival, the refugees are kept in undisclosed transit camps. Many refugees spoke to the participants regarding the refugee status and valid identification. According to UNHCR's global guidelines, any refugee claimant may appeal a negative decision concerning refugee status. "We don't feel like humans here and my younger brother who is born in the camp is even in a worst situation as he has no rights at all," said Tahshi, a young Tibetan teenager in the camp.

To get to Nepal the refugees have to pass from side to side the snow-covered Himalayas, the world's highest mountain range. Many among the fleeing peoples fall prey to the cold and freeze to death. Dekhi Lama, a young Tibetan refugee remembers a 7-year old girl who died of hypothermia and difficulty of breathing at high altitudes and was buried in the snow. This is one such example. Once settled in Nepal, some Tibetan refugees decide to stay while others choose to move on to India and other parts of the world.

Though the UNHCR must inform the applicant why someone's application has been rejected, the claimants were not given any reasons for why their applications were refused. Nearly 300 newly arrived Tibetan refugees were sent back. They were handed over to the Chinese authorities by the Communist Party of Nepal (CPN) in 1995, though the Foreign Minister Prakash Chandra Lohani assured that the Tibetan refugees need fear such consequences anymore. Eight of the Tibetans, aged 14 to 18 had been forcibly removed from Honuman Dhoka jail by the Nepalese and Chinese police and officials at 7:30 am on May 31, 2003. This deportation was implemented despite strong international lobbying in support of the Tibetans over few days and high level interest. Sometimes Nepalese security guards demand bribes from the refugees caught at the border, threatening to hand them over to Chinese authorities. Sometimes the refugees were given prison sentences ranging from seven to ten months as they were not able to pay the fines. Invariably the refugees in the prison are denied access to doctors, even if they are sent by the Nepalese. In 2000, Nepalese establishment and UNHCR agreed to change the existing one-page travel document for the Tibetan refugees to a passport style document that will be accepted by most countries. The new passport system was about to be implemented beginning in January 2001. On May 2003, the Government of Nepal turned 18 Tibetan asylum seekers including minors, over to the Chinese representatives to be forcibly repatriated to China. The actions taken against the Tibetan refugees not only violates international norms and practices regarding the humane treatment of asylum seekers and refugees, but also puts a question mark on Nepal's long standing reputation as a peace loving country with tolerance and hospitality.

TIBET: NEPAL URGED NOT TO RETURN REFUGEES TO CHINA

The Nepalese delegation was responding to the U.S. delegation statement urging them to continue "its long-standing commitment to ensure the safe passage through Nepal of Tibetan asylum seekers and refraining from forcibly returning Tibetan asylum seekers to China." "While we appreciate that the U.S. Administration had raised the plight of Tibetan refugees in Nepal, we remain shocked at the total denial by the Nepalese delegation over the refoulement of Tibetan refugees when cases have been documented by the United Nations High Commissioner for Refugees (UNHCR), Governments, NGOs and the media," Ms. Tsering Jampa, Executive Director of ICT-Europe said.

Although the Nepalese delegation was led by Deputy Prime Minister and Foreign Minister Ms. Sujata Koirala, it was the Nepalese Ambassador who told the UN mechanism: "On the issue of Tibetan refugees, Nepal stated that it does not allow its territory to be used against neighbours or any country. Anyone found in violation of the immigration law or other laws is dealt with according to the law of the land. There are no forceful deportations." In an advance written question, Switzerland stated: "What measures does the government plan to take, in order to ensure that the status of Tibetan refugees in Nepal is recognized according to international standards?"

As the draft report on Nepal was adopted this afternoon by the Council's Working Group on the Universal Periodic Review (UPR), Nepalese refused to accept the recommendation of the U.S. delegation to: "Protect vulnerable refugee populations by allowing for registration of the refugee population in Nepal and by refraining from forcibly returning Tibetan asylum seekers to China:" In a way to explain their position, Nepal changed its tone by again claiming that "there was no policy of forcibly returning the refugees." It should be recalled that on 31 May, 2003, the UNHCR in a press statement expressed grave concern about the fate of 18 Tibetan asylum seekers deported to China by Nepal. The statement said that the Nepalese authorities "decided to go ahead with the deportation on Saturday, ignoring last-minute appeals by the UN refugee agency and human rights groups. UNHCR learned that the deported group included eight unaccompanied minors, some as young as 13." UNHCR officials described the move as a blatant violation of Nepal's obligations under international law.

As documented by our annual Dangerous Crossing reports, ICT's written submission to the UPR mechanism "urged Nepal to uphold the principle of non-refoulement by taking adequate policy and administrative steps, including written policy instruction to immigration officials and border police, and training of Nepali police, security forces and immigration authorities about proper procedures and international human rights norms." Nepal also rejected recommendation by Algeria, Moldova, Netherlands, Slovenia and Switzerland who urged the country to become a Party to the 1951 Convention relating to the status of Refugees and its 1967 Protocol. Similarly the call from Netherlands that Nepal adopts national legislation pertaining to refugees that includes the rights of refugees and asylum seekers was not accepted.

ICT's submission expressed concern at the failure of establishment to issue refugee identity certificates to all refugees who met the criteria set by the Government. Thousands of eligible refugees had been waiting for years for processing of the certificates to resume. ICT called for durable solutions for long-staying refugees. Concern was also expressed at the practice of pre-emptive detentions and the wide-scale use of security forces used to intimidate Tibetans residing in Nepal and called on Nepali officials to ensure the fair treatment of Tibetans under Nepali law.

On 13 April, 2007, the UN Committee Against Torture after reviewing Nepal's second report said that the Committee was "disturbed about allegations received concerning cases of refoulement of Tibetan asylum-seekers, given the absolute nature of the prohibition against refoulement" under article 3 of the Convention Against Torture. The Committee also recommended that Nepal "enact legislation aimed at prohibiting refoulement of persons without an appropriate legal procedure...provide to the Committee information on the number of cases of extradition, removal, deportation, forced return and expulsion that have occurred since 1994, as well as information on cases in which deportation was not effected for fear of torture." On the overall performance of the Nepalese delegation at the UPR examination, the Nepal NGO Coalition for the UPR, representing 235 human rights and civil society organization said in a statement on 25 January that they were disappointed "at the rhetorical statement by the government delegation and their failure to provide any concrete commitments and timelines for the implementation of Nepal's human rights obligations."

The statement added that the Coalition "is troubled by the response of the government delegation who claimed that there is no systematic torture in Nepal, in spite of well documented and credible reports of systematic practices of torture at the hands of the State security forces." According to the Human Rights Council, by the end of the current two-week session of the Working Group on the UPR, 159 States (out of 192 Member States of the United Nations) will have had their human rights records examined and in December 2011 all United Nations' Member States will have been reviewed under this mechanism.

The UPR was created through the UN General Assembly on 15 March 2006 by resolution 60/251, which established the Human Rights Council itself. It is a cooperative process which, by 2011, will have reviewed the human rights records of every country. Currently, no other universal mechanism of this kind exists. The UPR is one of the key elements of the new Council which reminds States of their responsibility to fully respect and implement all human rights and fundamental freedoms. The ultimate aim of this new mechanism is to improve the human rights situation in all countries and address human rights violations wherever they occur. By the end of the current two-week session of the Working Group on the UPR, 159 States (out of 192 Member

States) will have had their human rights records examined and in December 2011 all United Nations' Member States will have been reviewed under this mechanism.

ORIGINS: TIBETAN REFUGEES RESIDE IN NEPAL

Nowadays, approximately 20,000 Tibetans reside in Nepal. The vast majority of this population arrived before 1989 and particularly during the period between 1959, when the People's Liberation Army's (PLA) assault on Lhasa sent thousands of Tibetans into exile, and 1974, when the Nepalese government forcibly shut down the Tibetan guerilla base that had been operating for more than fifteen years in the western Kingdom of Mustang. The remainder of the Tibetan refugee population in Nepal is comprised largely of the children of this first caseload and, to a much lesser extent, Tibetans who have fled since the late 1980s. The latter group, the second caseload, almost invariably proceeds to India because the Nepalese government no longer provides sanctuary to newly arriving Tibetan refugees.

In journey from the 1959 Uprising

While some Tibetan refugees arrived in Nepal in the early 1950s, the first major influx crossed the border in 1959, following the Lhasa Uprising. At that time most did not expect to remain in Nepal for more than a few months. They established camps primarily in the Himalayan border regions of Nepal such as Mustang, a small, largely autonomous, and ethnically Tibetan kingdom in western Nepal, as well as in Nubri, and Solu Khumbu. The climate in these regions resembles that of Tibet.

The refugees also felt that the proximity of the camps to the border would enable them to return home easily when the appropriate time came. Shortly, however, it became apparent to many that this time would not arrive in the near future. But few possessed the means to establish settlements with long-term economic viability. By 1961, many of the refugees faced serious food shortages and suffered from a lack of adequate shelter and healthcare.

In May 1960, Nepal requested assistance from the International Committee of the Red Cross (ICRC) and other aid organizations. Funded primarily by the United States Agency for International Development (USAID), the ICRC established emergency relief programs for the refugees. With airplanes provided by the Swiss Red Cross, the ICRC parachuted supplies to refugee encampments in the remote Himalayan regions. At about the same time, Tibetan officials of the government-in-exile returned to Dharamsala, India, from a fact-finding mission to Nepal, and reported to the Dalai Lama on the dire conditions facing the refugees there. In response, the Tibetan government established the Kathmandu Tibetan Welfare Office, a branch of the Tibetan Ministry of Home Affairs. The Tibetan Welfare Office functioned as a liaison between the refugees, various aid organizations, and the Nepalese government. Also known as the Office of Tibet, it continues to serve this vital role today.

The Guerilla process at Mustang

The guerilla operation in Mustang created a second major source of Nepal's original Tibetan refugee population. From about 1959 to 1974, the Nepalese government turned a blind eye to a Tibetan military base located in the remote northwestern Kingdom of Mustang, which received covert support from the United States Central Intelligence Agency (CIA). From this base, Tibetans, primarily former residents of the province of Kham, carried out guerilla strikes against the PLA. For about ten years, the CIA funded military training for Tibetan guerillas, including in "camouflage, spy photography, guns, and radio operation," and provided them with weapons and other equipment to sustain the Mustang base. By the late 1960s, however, the U.S. government's support for this covert operation dwindled, and the CIA gradually discontinued aid to the guerillas. At about the same time, Nepal began to seek a closer relationship with China because "[n]o longer fearful, as it had been in the early 1960s, of a Chinese attack, Nepal now wished to counter New Delhi's influence in the region by furthering ties with [Beijing]."

In 1974, the Nepalese government ordered the remaining Tibetan guerillas (about 1800 men) to shut down the operation. To avoid violent conflict with the Nepalese and the endangerment of the Tibetan refugee community in Nepal, the Tibetan administration in Dharamsala agreed. The Dalai Lama then conveyed a personal, tape-recorded message to the guerilla forces requesting them to lay down their arms. The disbandment of the Mustang guerilla forces created a second population of Tibetan refugees that required resettlement. Trained for war, however, this group, like those who fled in the aftermath of the 1959 Uprising, did not possess the knowledge or means to establish new settlements with long-term economic viability. These two groups comprised the first caseload of Tibetan refugees in Nepal and established the twelve major Tibetan refugee settlements.

Establishment and growth of the Settlements

In the early 1960s, the Nepalese government arranged to provide the first caseload of Tibetan refugees with land. It established four "temporary" settlements: (1) Chialsa, in the Solu Khumbu mountain range east of

Kathmandu; (2) Tashi Palkhiel, on the outskirts of Pokhara; (3) Dhorpatan, in western Nepal; and (4) Jawalakhel, on the southern edge of Kathmandu. The Nepal Red Cross (NRC), founded in 1963, purchased the land for these settlements with funds donated by UNHCR. Apart from their inability to afford the land at that time, Tibetan refugees did not enjoy the legal right to purchase it. They required an intermediary to hold the land in trust for their use. The NRC served this critical function. It also enabled the Nepalese government to disavow any direct support for the refugees in order to avoid jeopardizing its tenuous relationship with China. The result, as the NRC's present executive director remarked, was that the NRC aided the refugees "with the [tacit] *support* of the government," but "not exactly *for* the government." By 1969, seven Tibetan settlements had been established in Nepal.

As it became clear the their exile would not be short-term, the Tibetan government required to bring the refugees scattered throughout Nepal together in the newly established settlements. Shanmo Palchung, an elder Tibetan at the Tashi Ling Settlement in Pokhara, recalled that in 1964 Tibetan officials asked him to gather the refugees temporarily settled at encampments near the border and bring them to Pokhara. He arrived in Pokhara that year with 377 Tibetan refugees from throughout Nepal and then petitioned the NRC to purchase land for them from the Nepalese government. The Nepal Red Cross "gave [us] land for living and agriculture," he said, "but it was bad [and undeveloped] land"; and in the first year, sixty refugees, living without shelter, died from exposure to the heat and monsoon rains. The Nepalese government soon recognized that it must take steps to make the refugee settlements self-sufficient.

In 1964, Nepal asked the Swiss government—acting through the Swiss Red Cross and the Swiss Association for Technical Assistance—to help to resettle the refugees and administer the settlements. On November 22, 1964, the two governments entered into an agreement stipulating the main goals of the settlements: long-term economic self-sufficiency and gradual integration into Nepalese society. Thereafter, the Swiss groups provided the residents of the settlements with basic humanitarian aid, technical assistance, and vocational training. Other organizations, including USAID, the Protestant United Mission, the Nepal International Tibetan Refugee Relief Committee, the Norwegian Refugee Council, and United Nations affiliates, such as the U.N. Children's Fund (UNICEF) and the World Health Organization (WHO), contributed additional aid in the form of medical care, primary school education, construction of housing, and food rations. At about the same time, the Tibetan government-in-exile recognized a need to establish its own permanent representatives to administer the settlements. For each settlement, it appointed a "Welfare Officer." These individuals cooperated with the Tibetan government's principal representative in Kathmandu to administer the settlements and look after the needs of their residents.

UNHCR offered monetary assistance to the ICRC and other aid organizations from the outset. Its role expanded significantly, however, in 1964. At that time, the High Commissioner, at the request of the Nepalese government, sent an envoy, J.D.R. Kelly, to appraise the situation. Kelly reported that Tibetan refugees remained in dire need of humanitarian assistance. In August 1964, the Nepalese government authorized UNHCR to open an office in Kathmandu to serve as a liaison between the Tibetan community and the government. In 1965, the government asked UNHCR to begin to assist with the resettlement of those Tibetans not already residing in one of the initial camps. UNHCR remained actively involved throughout the 1960s and the early 1970s. Its role shifted over time.

Initially, UNHCR focused on providing urgently needed emergency relief; in later years it began to search for more permanent solutions. In 1973, however, UNHCR discontinued its assistance program for Tibetan refugees and closed its Kathmandu office. It concluded that Nepal was well on its way to achieving a durable solution to the problem through local integration. In the early years, Tibetans worked to make sure the long-term availability of basic necessities for the settlements. With the assistance of non-governmental organizations, they built roads, homes, schools, and medical facilities. Thereafter, each settlement began to establish industries with which to sustain itself. These included, for example, agriculture, handicraft production, and carpentry. But the refugees achieved their greatest financial success in the carpet industry. Swiss aid workers first encouraged the development of this trade at the Jawalakhel, Chialsa, and Tashi Palkhiel settlements. In 1966, the factories were converted into private companies managed, though not owned, by the Tibetan refugees themselves. By the early 1980s, the Tibetan carpet industry had "evolved into the primary source of income for the entire Tibetan community in Nepal." By the late 1980s, it had become the largest handicraft industry and the second-largest earner of foreign currency in Nepal. Some Tibetans also recognized private shops, hotels, and restaurants, particularly during the first two decades of their exile. Until the late 1970s, Tibetans in Nepal, by contrast to those in India, "did not need any legal document or down payment to open their stores." The mass exodus of Tibetans in the late 1960s and early 1970s also coincided with Nepal's economic expansion and the development of its tourist industry. This, too, created some opportunities for Tibetan entrepreneurs. Restrictions on their rights to own property, travel, register businesses, and work, however, prevented them from securing a strong foothold in the tourist industry.

Today, more than one dozen Tibetan settlements exist in Nepal, including Jawalakhel, Boudha, Swayambunath, and Jorpatia, in the vicinity of Kathmandu; Jampaling, Paljorling, Tashi Ling, and Tashi Palkhiel, in the Pokhara region; and Dhorpatan, Chialsa, Chairok, Shabrus, and Lumbini, in the northern regions of Nepal. The majority of these were established either in the early to mid-1960s or in 1974, when the Nepalese government terminated the Mustang guerilla operation. Six of the settlements rely on handicraft industries, four sustain themselves by agriculture, and the remainder are "self-reliant," meaning their residents work individually in various jobs.

Nepal, with the aid of foreign governments and confidential donors, thus succeeded in resolving the immediate humanitarian crisis provoked by the mass exodus of Tibetans in the aftermath of China's annexation of Tibet and the 1959 Lhasa Uprising. But the prospect of a long-term and durable solution that led UNHCR to close its Kathmandu office in 1973 has proven elusive. Because of the isolation of the settlements and their residents from mainstream Nepalese society, local integration of the Tibetan refugee population has been minimal. Tibetans who were resettled in the 1960s and 1970s are permitted to remain in Nepal, but their legal status is ambiguous and their financial situation precarious. Without the rights to own property and businesses, to travel, and to work freely in industries and vocations outside the settlements, Tibetans residing in the settlements remain economically and socially alienated from Nepalese society. With the notable exception of some Tibetans in the carpet industry, which has been in decline in recent years, relatively few of the refugees in the settlements have been able to improve their standard of living far beyond subsistence. The precarious status of the settlement residents highlights the need for a more durable solution that either permits the refugees to acquire Nepalese citizenship or affords them a more expansive set of rights consistent with those ordinarily afforded to permanent resident aliens under international law.

TIBETAN DIASPORA

The Tibetan diaspora is a term used to refer to the communities of Tibetan people living outside Tibet. Tibetan emigration happened in two waves: one in 1959 following the 14th Dalai Lama's self-exile in India, and the other in the 1980s when Tibet was opened to trade and tourism.

The third wave continues from 1996 to today. Not all emigration from Tibet is permanent; today some parents in Tibet send their children to communities in the diaspora to receive a traditional Tibetan education.

In the 2009 census, Tibetans in exile registered are about 128,000, with the most numerous part of the community living in India, Nepal, and Bhutan. However, in 2005 and 2009 an estimation at up to 150,000 was given.

Origins and Numbers

The Central Tibetan Administration (CTA) provides a Green Book-a kind of Tibetan identity certificate-to Tibetan refugees. Based on a CTA survey from 2009, 127,935 Tibetans were registered in the diaspora: in India 94,203; in Nepal 13,514; in Bhutan 1,298; and in rest of the world 18,920. However, their number is estimated at up to 150,000, as mentioned by both Edward J. Mills et al. in 2005 and by the 14th Dalai Lama in 2009.

The larger of the other communities are in the USA, Canada, the United Kingdom, Switzerland, Norway, France, Taiwan and Australia.

First Wave

During the 1959 Tibetan uprising, the 14th Dalai Lama and some of his government fled to India. From 1959 to 1960, about 80,000 Tibetans followed the Dalai Lama to India through the Himalayas. Continued flights, estimated in the numbers of 1,000 to 2,500 a year, increased these numbers to 100,000. The movement of refugees during this time is sometimes referred to as an "exodus", as in a United Nations General Assembly resolution in 1961 that asserted that the presence of Tibetan refugees in neighbouring countries was "evidence" of rights abuses in Tibet.

Second Wave

After the opening of Tibet in the 1980s to trade and tourism, a second Tibetan wave of exile took place due to increasing political repression. From 1986 to 1996, 25,000 Tibetans joined and increased by 18% their exiled community in India. This movement of refugees during this second wave is sometimes referred to as a "second exodus". According to a US cable put out by WikiLeaks, from 1980 to November 2009, 87,096 Tibetans arrived in India and registered at the Dharamsala reception centre, whereas 46,620 returned to Tibet after a pilgrimage in India. Most of those staying are children to attend Tibetan Children's Villages school.

Present Emigration

A 2008 documentary directed by Richard Martini claimed that 3,000–4,500 Tibetans arrive at Dharamsala every year. Most new immigrants are children who are sent to Tibetan cultural schools, sometimes with the tacit approval of the Chinese government. Many political activists, including monks, have also crossed over through Nepal to India. Significant cultural gaps exist between recent Tibetan emigrants and Indian-born Tibetans. The more established Tibetans in diaspora reject recent immigrant Tibetans who watch Chinese movies, sing Chinese music, and can speak Mandarin, who are then alienated from the exile community. Newcomers express frustration that the government-in-exile wants to hear only "bad things" about Chinese rule in Tibet, and a lack of economic opportunity in Dharamsala.

In India

Organisations

The main organisation of the Tibetan diaspora is the Central Tibetan Administration of the 14th Dalai Lama based in the McLeod Ganj suburb of the city of Dharamsala in India. One of the people to organize the functioning of Tibetan refugee camps in India and to transform the life of Tibetan refugees into the ways of a diaspora was Polish born Wanda Dynowska (1888–1971), called Umadevi, recalled by the 14th Dalai Lama as the "Polish Mother", along with her friend, Indian citizen of Polish origin, Maurycy Frydman. The CTA maintains Tibet Offices in 10 countries. These act as de facto embassies of the CTA, offices of culture and information and effectively provide a kind of consular help to Tibetans. They are based in New Delhi, India; New York, USA; Geneva, Switzerland; Tokyo, Japan; London, UK; Canberra, Australia; Paris, France; Moscow, Russia; Pretoria, South Africa; and Taipei, Taiwan. The Tibetan diaspora NGOs deal with the cultural and social life of the diaspora, the preservation of cultural heritage, and the promotion of political Tibetan independence.

Education

The Central Tibetan Schools Administration with a seat in New Delhi is an autonomous organization established in 1961 with the objective to establish, manage and assist schools in India for the education of Tibetan children living in India while preserving and promoting their culture and heritage.

According to information on its own website, as of 2009 the Administration was running 71 schools in the areas of concentration of Tibetan population, with about 10,000 students on the roll from pre-primary to class XII, and with 554 teaching staff. According to the information on the website of the CTA, as of 2009.01.13. there were 28 CTSA schools whose enrolment was 9,991 students.

In 2009, The Tibetan Children's Villages established the first Tibetan higher college in exile in Bangalore (India) which was named "The Dalai Lama Institute for Higher Education". The goals of this college is to teach Tibetan language and Tibetan culture, but also science, the arts, counselling and information technology.

MIGRATION FROM SETTLEMENTS IN INDIA

Migration of young people from Tibetan settlements in India is a serious cause of concern as it threatens Tibetan identity and culture in exile with marginalization. According to Tenzin Lekshay, most exile settlements are guarded by old aged people, some established schools in the settlements are on the verge of closing for lack of pupils, and graduates are scattering to Indian cities because of the lack of employment opportunities in the community. According to Nawang Thogmed, a CTA official, the most oft-cited problems for newly migrating Tibetans in India are the language barrier, their dislike for Indian food, and the warm climate, which makes Tibetan clothing uncomfortable. Some exiles also fear that their Tibetan culture is being diluted in India. Indian television runs in Hindi and English.

In Bhutan

Few Tibetans settled in Bhutan after 1959, as the country was used mainly as a transit route to India. However, in 1961, following growing tensions between China and India, India sealed its northern border with Bhutan, prompting Bhutan to arrange an emergency meeting with the Government of India (GOI) and the CTA to deal with the Tibetans stuck in the country.

The government of Bhutan agreed to take in 4000 settlers, although ordinary Bhutanese became increasingly resentful of the Tibetan immigrants because of their refusal to assimilate into Bhutanese culture.

In 1974, 28 Tibetans, including the representative of the Dalai Lama in Thimphu, were arrested and accused of a conspiracy to assassinate King Jigme Singye Wangchuck. When the CTA refused to provide evidence of their innocence, relations between Bhutan and Dharamshala soured, and in 1979, the Government of Bhutan announced that any Tibetan in the country that did not take Bhutanese citizenship would be repatriated back to China. Despite the CTA's opposition, 2300 Tibetans applied for citizenship; most of the remainder resettled in India.

In Europe and North America

Tibetans in Europe and North America are not as politically active or unintegrated into the host society as Tibetans in India, because they do not live in concentrated communities, and are educated in local schools instead of living under a special Tibetan jurisdiction as Tibetan Indians do. In Europe, Tibetans have been received more positively than other immigrants to Europe, and are stereotyped as spiritual and victims. Politically, organizations such as the Tibetan Youth Association in Europe attempt to cultivate a Tibetan identity for Tibetans in Europe. However, many Tibetan youths in Europe cannot speak a Tibetan language and do not feel they belong to a wider Tibetan community; such identity crises have on occasion led to drug dependency and crime.

WEAVING A FUTURE FOR TIBETAN REFUGEES

The recent history of Tibet has been one of mourning. Despite the relative impenetrability of its mountains, it was invaded and occupied by armies of the Communist Peoples Republic of China in 1950. After an unsuccessful popular revolt in 1959, thousands of Tibetan refugees fled across the Himalayan Mountains to India and Nepal, and a government in exile was established under the Dalai Lama, the traditional leader of Tibet. Since that time, the refugee communities have grown larger every year, as more Tibetans flee severe political and cultural repression in their Native land. Today, despite one of the most successful resettlement programs in modern history, the needs of over 500,000 refugees exceed available resources. In response to these events Cultural Survival established the Tibetan Rug Weaving Project.

The project intends to maintain in one collection rugs that represent the various sources of inspiration found in fine antique Tibetan carpets, while at the same time drawing on sources of inspiration from throughout the traditional and modern world. All rugs woven in the project have colours derived entirely from customary vegetal dyes. Other than indigo (blue), which is obtained in India, these dyes are indigenous to Nepal. Dye sources include madder root (various shades of red), walnut husks (dark brown), several different local flowers and roots (yellows), and an overlay of indigo on yarn previously dyed yellow (green). Of course, the wool is as important as the dyes. Only pure handcarded and handspun Tibetan wool is used in project rugs. Carpets woven with it are not only extremely long-wearing, but also have a luster that increases with use.

The Cultural Survival Tibetan Rug Weaving Project was founded in 1990. Its original goals were to explore the rich heritage of the Tibetan weaving tradition and to generate funds to benefit the Tibetan community in exile. As this project enters its second decade we look back on an achievement we could not have imagined.

The project has worked with the Srongsten School, a Tibetan exile government school in Nepal, for 10 years, and sponsors the education, rooming, and board of more than 70 Tibetan refugee students there. It has built two schools in India for the exile government Department of Education, one in Puruwalla, a small settlement in Himachal Pradesh, and a second in Pokrebeong, a hill settlement in eastern India near Darjeeling. The project also sponsors food supplement programs in five exile government schools, four in eastern India, and one at Sarnath near Varanasi.

It has also sponsored the shipment of two containers of textbooks from the United States to the Tibetan exile government schools in Nepal and India, and has funded the construction of a computer classroom in Nepal. In 1998 the Tibetan weave Project took its first step toward working inside Tibet with the initiation of a reforestation project in the Potamo region of Kham in eastern Tibet, which has been denuded of trees by clear-cutting. The project has transplanted more than 400,000 seedlings to the hillsides in that region.

It has also established extensive fruit orchards throughout the region and has begun a pilot program to train local youth to identify, process, and use traditional Tibetan medicinal plants. In July 2002 the project completed its largest undertaking to date with the inauguration of a school in the Litang district of Kham in eastern Tibet, funded by rug sales. The school was the first in the remote Chungba Valley. It has classrooms for 210 students, boarding facilities for students and teachers, and a large kitchen and dining hall. The school was built entirely of wood in a traditional Tibetan style and will be a Tibetan medium school with a Tibetan principal and all Tibetan teachers, except one who will teach Chinese language.

The project's first rugs concentrated primarily on Tibetan designs, but the project's participants realized that the market was much too diverse to restrict themselves in this way. The weavers now continue to weave traditional designs, but the inventory of over 250 designs contains those inspired from traditions in every part of the world as well as modern designs of weavers' own creation. The weavers also have a sub-project called Gaon Naksha for which they create their own designs as they weave. These carpets are free-form, spontaneous expressions of the weavers' life experiences, usually portraying their origins in villages of Nepal.

TIBETAN REFUGEE CLAIMANTS IN TORONTO

The global prevalence of tuberculosis (TB) infection is predictable to be 32%, with approximately 8 million new symptomatic cases and 1.9 million TB-related deaths annually and a rising incidence of drug-resistant TB (DRTB).

Ten percent of new cases of active TB are resistant to at least one first-line agent, and 1% have multidrug-resistant TB (MDRTB) [resistant to at least isoniazid and rifampin]. The annual incidence of new cases varies from > 250 cases per 100,000 in some heavy-burden nations, to < 10 per 100,000 in low-burden countries (including Canada and the United States). Drug-resistance rates are highest in places such as Russia, where rates of DRTB and MDRTB are 30 to 40% and 10%, respectively. Canadian figures correspond to the global median, 10% and 1%, respectively, with slowly increasing levels over the past decade in the province of Ontario.

As the incidence of TB in Canada dropped in the second half of the 20th century, the proportion of cases diagnosed in foreign-born individuals rose steadily, reaching nearly 60% by 1995 and 85% in Ontario. This reflects migration patterns from highly TB endemic countries. Current Canadian screening policy for refugee claimants in Canada requires the initiation of appropriate investigations for TB within 30 days of arrival if their claim was made outside of Canada, or within 60 days of their claim for applicants within Canada. This system could lead to delayed or incomplete screening.

Very high rates of vigorous and MDRTB were reported in a group of Tibetans from India and Nepal who entered the United States in 1990. No publications on subsequent migration of this or similar groups have been reported. Between 1998 and 2000, a group of Tibetan people who had entered the United States as visitors entered Canada and claimed refugee status. Beginning in 1990, as part of a humanitarian effort, there was a resettlement of Tibetans from India and Nepal to the United States. Some Tibetans who initially came to the United States subsequently entered Canada and claimed refugee status. Members of our cohort sought refugee status in Canada after lengthy US stays as visitors, because they were no longer able to maintain their visitor status or otherwise extend their stay in the United States. Most initially settled in Toronto, and many had TB screening completed at our centre. We sought to review TB-related outcomes in this cohort, to compare our experience with the findings of Truong et al, and to assess longer-term follow-up after a stay in a low-incidence region.

VIEWING COMMUNITY OUTSIDE THE HOMELAND

In both cases there is definitely a trend toward the unification of sectarian differences. In Judaism, the rabbis seem adamantly against sectarian controversy and develop a whole style of literature that bases itself on including different opinions. Meanwhile, the exiled Tibetan orders have developed ecumenical meetings and discussions aimed specifically at better relations between the groups as well as creating an inclusive governmental structure within the newly developed constitution. Nevertheless, there is a distinct difference between the two cases. While in the Jewish case there is a move from exclusivist claims to more pluralistic ideals, such pluralism already existed in pre-exilic Tibet. Even in pre-exilic times, Tibetan Buddhists recognized that there can be multiple paths to one goal, that there can be differences of opinion without necessitating that one be "more right" than the others.

Then again, there is the case of Bon, which in pre-exilic Tibetan Buddhism was not given the status of a legitimate path, and at times, was even persecuted. In exile, relations with the Bonpos in particular have improved, to the point where there is significantly less hostility, more respect given for each other's systems, and at the extreme, Bon is referred to as the fifth order of Tibetan Buddhism. Suddenly, a relationship begins to emerge between the changes made to sacred space in exile and the changes made to sacred community in exile. The singular sacred space of the temple in Jerusalem becomes multiple in exile, as God's presence—the shekhinah— descends upon any synagogue or any group of Jews studying Torah, although the presence of God was still rationalized as "one." Meanwhile, the sacred pilgrimage spaces in pre-exilic Tibetan Buddhism were multiple, and there was little tension in the idea that they became even more numerous in exile, incorporating new sites in India and Nepal, including any site the Dalai lama visits, because the theology is already multiple in Buddhism, to an extent.

Now we find a similar situation with regard to sacred people, sacred community, "Israel" for Jews, and "insiders" for Buddhists. The exclusivist claims to being the people "Israel" become more open in exile, whereas the more inclusive "insiders" of Tibetan Buddhism, while becoming more inclusive of Bon and while improving relations among the sects, do not need the same rationalization, because the pre-exilic societal structure, as well as the theology, were already pluralistic. The more we look into this, the more we see that it is not so much what Tibetans can learn from the Jews as it is that the Jews had to change to accommodate exile, whereas Tibetan Buddhism was already structured to adapt well to the situation of exile.

Yes, Tibetans had regional, theological, and political groupings in pre-exilic times and have been changing the power structures in exile to reaffirm the affinity of being Tibetans over the previous estrangements. And certainly, as a result, there has been some of the kind of unification that took place in Judaism, as well as a Gelugpa-ization that may be similar to the Pharisization that took place in Judaism. But unlike in Judaism, the reduction of estrangement in Tibetan Buddhism does not erase the earlier affinities. There is a real effort to maintain regional and theological identities: the question of whether or not this will be ultimately successful will depend on the length of time of the exile as well as the continued effort to instill the next generation with different ideas.

The question of whether or not new affinities would be formed was the intended focus of this chapter, as well as the hypothesis, given the psychological and political needs of refugees, to present themselves as a unified force. What then, do we make of the new estrangements? The advent of the Shugden controversy creates a problem for this hypothesis. Rather than promoting affinities, these groups are creating sentiments of estrangement. The heavy involvement of the Chinese government in Shugden demonstrates a political reason to create estrangements in exile. And the involvement of Europeans complicates things further. Perhaps one could view this sect in the following way: Shugden is not solely concerned with estrangement from, but also with affinity for other groups, whether Western or Chinese.

This brings us back to the Jewish case. Do the early Christians fit the same pattern? In the Christian case, the answer is assuredly "yes." Considering synagogues housing both Jews and Jewish-Christians existed possibly up through the fourth century, Christianity as separate from Judaism was not built entirely on a discourse of estrangement, but also on an affinity for other non-Jewish groups. Looking at groups this way, one might consider splinter groups that did not survive, such as the followers of Bar Kochbah. They used a discourse of estrangement as they turned to a belief in Bar Kochbah as the messiah, but the sect dissipated shortly afterward. Perhaps the problem was not that the theology could not survive Bar Kochba's demise, but rather that the discourse of estrangement did not simultaneously create a discourse of affinity with some other group, such as the Christians with the Romans or the Shugden sect with the Chinese government or European supporters.

Along the same line of thought, one might consider another current controversy within Tibetan Buddhism, that of the two Karmapas. When the sixteenth Karmapa passed away, the search for the seventeenth Karmapa created two rival camps. Currently, the Karmapa born in Tibet has most of the support, although the Sharmapa, the main supporter of the Karmapa born in India, has been culling support, partially from abroad. Whether this splinter group will turn out to be a footnote in history or a serious faction will perhaps not be determined by finding a way to reconcile the theology of the matter, but rather by finding the appropriate regional/and or political group with which to align itself. Yet again the prominence of belief over other factors comes into question.

This essay has intended to show that sentiments of estrangement are often replaced with those of affinity for a larger grouping in exile due to the need to survive when confronted with new surroundings. Both case studies exemplify these new discourses of affinity, and show that new estrangements are also possible, if attached to a discourse of affinity with an outside group. These discourses of affinity, whether external or internal, need not necessarily be with regard to belief, but can be political, regional, or even economic.

However, turning back to the more prominent idea of unification, we find a difference in the two cases. Unlike the Jewish case, in which new affinities are intended to erase sentiments of estrangement, the Tibetan case shows a new possibility: that of forging new relationships of affinity while maintaining separate identities. Clearly the time periods we are looking at are a factor here. In the current global situation and given the current discourses of pluralism and multiculturalism, the idea of grander affinities while maintaining subsets of distinct affinities has become a real possibility. This also reflects a difference in the degree of estrangement felt in pre-exilic times.

The Tibetan groups were not sectarians claiming sole knowledge of the truth. While there were problems between regions and theological orders, an overarching acceptance of different paths leading to one goal kept the situation from being one where a new affinity would contradict an old affinity. In Judaism, to become a rabbinic Jew was to become a universal Jew and give up any previous affiliation with a particular subgroup. There are no Essenes, Sadducees, or Pharisees left in rabbinic times, nor are there northerners or southerners. There are of course Palestinians and Babylonians, but even this distinction becomes minimal in the new discourse of the Mishnah and Talmud and in the methods used to evoke sentiments of affinity between the two groups. Posman: Post-Exilic Reconstructions of Sacred Community 30

What does all this mean for the future of Tibetan society? That is still unsure. There is ample evidence to suggest that each order will survive with its own prayer festival, lineage histories, political representatives, teachings on emptiness, and meditative and ritual practices. However, as exile wears on, the more likely outcome is that the ecumenicism between the orders will result in an eventual single order that will unify the populace; it will be largely Gelugpa and largely descendant of Gelugpa, but by that time, aspects of the other orders will imbue the sole order as well.

As for regional issues, the process will be quicker. Already second generation refugees are confused about what region to vote with and are largely unaware of the dialect, customs, clothing, and landscape of their home province. Should such refugees return, it is doubtful that they would return to a home province. Exhibits and cultural productions will remain a major form of conveying this knowledge to future generations, but pre-exilic regional identities seem to dissipate faster than theological ones, while new regional affinities by settlement or even country of residence are also emerging more quickly than any other type of affiliation.

TREATMENT OF THE REFUGEES

In the beginning, two large transit camps were set up in consultation with the Government of India: one at Missamari, located ten miles from Tezpur in the Indian State of Arunachal Pradesh; the other at Buxa Duar, a former British prisoner-of-war camp situated near the Bhutanese border in West Bengal. Within a few weeks after the setting up the camps, 6,000 refugees arrived at Missamari and 1,000 at Buxa Duar. As the camps could accommodate not more than 9,000 persons between themselves, simultaneous efforts were made to disperse the refugees for road building works in the cooler regions of North India to prevent fatalities due to intense heat, overcrowding, and epidemic of amoebic dysentery. The first batch of 3,394 people was sent for roadwork in Sikkim in September 1959 in several groups.

Agricultural base Settlements

The Dalai Lama's vision of exile society took root in his new headquarters, an abandoned British hill station called Dharamsala. On 29 April 1960, after a little more than a year's stay, the Dalai Lama moved to his new headquarters from Mussoorie. In Dharamsala, the Dalai Lama revitalized his small administration in exile and called a special meeting of the people's deputies and staff members on 2 September 1960 to formulate new strategy and plan. As a part of the new plan, he established a bureau (now known as Bureau of H.H. the Dalai Lama) in New Delhi to serve as the Tibetan link with the Indian Government and various international relief agencies that were coming to the refugees' aid. Similarly, the Office of Tibet in New York was also set up in 1964 to raise the issue of Tibet at the United Nations and to disseminate information on Tibetans inside and outside Tibet. Rehabilitation, however, was the instant task at hand.

Although the roadwork had provided the much-needed temporary employment for the refugees, it turned out to be not only dangerous, but also hard. The migratory and unsettled nature of the roadwork proved to be extremely hard, particularly on the children and old people. Apart from this, many road-workers suffered from tuberculosis and other communicable diseases due to lack of nutrition and sanitation, compounded by hard labour.

As such, the Dalai Lama approached Prime Minister Nehru for assistance for the permanent rehabilitation of the refugees. Nehru, on his part, canvassed the State Governments for vacant countryside for the purpose. Karnataka State (in South India) immediately responded, offering initially 3,000 acres of uninhabited stretch of jungle in gently rolling hills 52 miles west of Mysore City, situated at the altitude of 2,6000 ft. above sea level. It agreed to accept initially 3,000 Tibetan refugees and provide a grant of four million Indian rupees to execute the rehabilitation project.

Soon after this, the first Tibetan agricultural resolution was set up in the middle of December 1960 at Bylakuppe, accommodating a total of 3,217 Tibetan refugees by the end of 1965. For administrative reasons, the settlement was divided into six camps, each with 500 acres. In the first few years, the settlers engaged in reclaiming land, digging wells, and constructing houses, during which they were paid daily wages and given regular rations. The settlers were divided into artificial families of five people, relatives or not, and each family was given five acres of agricultural land. They were accommodated in the three-room brick-walled, tile-roofed single story house, built on a 500-sq. ft., part of which also served as their kitchen garden. A cooperative society was set up in 1961 and later formally registered. The cooperative society procured and advanced seeds, fertilizers, and pesticides; marketed agricultural produce collectively; and provided common services such as tractors, trucks, and warehousing. Besides, the cooperative society later played crucial role in managing carpet weaving centres, flour mills, poultry farms and other ventures such as retail outlets in the settlements. In the same year, a school was opened in the settlement. When the Bylakuppee settlement was first started, a small dispensary was set up in a tent to cater to the emergency medical needs of the refugees. Later in 1964, a special complex of concrete houses was built with assistance from foreign organizations to accommodate a dispensary with inpatient facilities.

For the first five years, the settlers cultivated cotton, tobacco, rice, millet, paddy, lentils and oilseeds, all of which failed miserably. Farming in South India was so different from Tibet's unique high-altitude agronomy that many settlers became totally discouraged and helpless. However, the agony of the repeated crop-failure was solved in 1965, when a Swiss technical adviser, in consultation with the Agriculture University of Bangalore, conducted soil tests and also tested the yields of bullock-ploughed and tractor-ploughed fields. This test resulted in the cultivation of maize crop instead of cotton, tobacco, and other crops and also replacement of bullock plows by tractor plows. The new crop and methods proved so successful that, by the end of 1966, the settlement became self-supporting. Bylakuppe settlement had served as a model for planning other agriculture based settlements later. By mid-seventies, 26 agricultural-based settlements were established in India, Nepal and Bhutan, accommodating a majority of the Tibetan refugees.

Agro-Industrial Settlement

It was apparent that it was not viable to restore all the refugees in agriculture-based settlements. Not only was it hard to find relatively large tracts of land, it was also not practical to rehabilitate all the refugees in agriculture-based settlements. Many of the refugees were traditionally nomads and traders, and had no idea of farming. With this in view and also to expedite the process of rehabilitation, it was found necessary to seek other avenues of resettlement.

Thus, a plan was made to rehabilitate many of the refugees in small-scale industries. To facilitate this, the Tibetan Industrial Rehabilitation Society (TIRS) was established in October 1965, with its headquarters in New Delhi. For the next three years, the TIRS launched several small industrial projects. These included woolen mill at Bir, tea estate in Bir and Chauntra, Tibetan crafts at Bejnath, limestone quarry at Kumrao, hydrated lime plant at Sataun, fiberglass factory at Poanta Sahib, and handicraft centre in Puruwala. These projects accommodated more than 4,000 refugees. These projects were meant not only to provide adequate employment opportunities to the inmates, but also skills in different trades. In each industrial project, school and health care facilities were provided by the TIRS with funds raised from foreign agencies. Besides, the TIRS established a Himalayan Marketing Society, based in New Delhi, to market the produce of these settlements. However, many of these industrial efforts failed due to lack of technical and managerial skills. The majority of the industrial settlements gradually began to depend more on traditional carpet weaving, roadwork in summer, and woolen garment trade in Indian towns and cities in winter.

Handicraft-based Settlements

Over the centuries, Tibet had shared intimate cultural and deal relationship with its neighbouring countries. Through these interactions resulted in the development of many traditional crafts in Tibet. The traditional crafts included metal work, woodcraft, clay modelling, applique work, thangka painting, hand-made paper, leather work, woolen cloth and carpet weaving, etc. Although there were no major institutes in the traditional society to promote these trades, these arts and crafts were practiced in individual households and the products were sold out locally, or through traders to different part of Tibet.

When the Tibetan refugees came into exile in early 1960s, some of the Tibetan craftsmen settled in Kalimpong, traditional Tibetan trading post in the North India, and others settled temporarily in Darjeeling and Nepal. Thus, with hope of preserving and promoting the traditional Tibetan crafts as well as to provide employment options for the refugees, the CTA, in consultation with the Government of India, set up several handicraft-based settlements as a part of rehabilitation scheme.

The first handicraft-based settlement was set up on a four-acre estate in Darjeeling in October 1959. The 13th Dalai Lama, during his exile from Tibet in 1910-11, had used the site as his temporary residence and therefore, has a special sentimental value for the Tibetans. The settlement or centre provided shelter and employment for 450 people. Education and medical needs of the inmates were also taken care of. Besides, the centre also ran old people's home and a crèche for small children.

The centre's produce, such as carpets, woolen sweaters, Tibetan women's aprons, wood carved decoration items, wooden toys, leather goods, etc. were marketed locally as well as abroad with the help of aid agencies. Encouraged by the success of the first handicraft-based settlement, other similar settlements or centres were set up in the following years. These included the handicraft centre in Dalhousie, Dharamsala, Kalimpong, Simla, Clement Town, Rajpur, and Dehradun–all in India. Similar settlements were also set up in Nepal. In these handicraft-based settlements, cooperative units were established to procure raw materials and market produce. While the cooperatives marketed the settlement's produce both in India and abroad, the settlers engaged also in individual trade of sweaters–knitted initially by themselves, but later bought from the factories in Ludhiana City in Punjab, India–to supplement their income. Each winter they set up makeshift bamboo-walled and plastic-topped stalls on the sidewalks of bazaars and marketplaces all across India. By the later half of the sixties, sweater selling became the supplementary trade of the majority of Tibetan refugees.

By 1980, only 30,000 Tibetan refugees remained on road construction work. By 1992, a total of 52, 500 people were rehabilitated in the agricultural settlements, 7,700 in Agro-Industrial settlements, and 6,400 in handicraft-based settlements, and 39,800 refugee remained outside the organized units, earning livelihood in petty private enterprises or in the Indian service industries.

New Challenges

The death of Mao Zedong in 1976 and the end of the Cultural Revolution in China and Tibet resulted in opening China to the outside world. This dramatic change also led to the opening of Tibet and the first-post 1959 contact between Beijing and Dharamsala, which took place in March 1979. Many Tibetans, both inside and outside Tibet, took the advantage of the situation to undertake the arduous cross-Himalayan journey to visit their family members and relatives. Many families met for the first time after two decades of separation. With the passage of time, a steady stream of new refugees from Tibet came to seek asylum in India and Nepal.

From 1987 to 1994, Tibet was rocked by a series of pro-independence demonstrations. The Chinese authorities reacted by escalating the machinery of repression, especially against those involved in demonstration. This made life in Tibet all the more difficult, resulting in a mammoth increase in the flow of refugees. Between 1986-1996, some 18,700 refugees arrived in India and Nepal. The situation became still worse after 1994 when the Chinese government introduced harsher policy to deal with the Tibetan nationalism in Tibet.

In order to cater to the needs of these new refugees, the CTA opened a Reception Centre in Dharamsala in 1990. Later, its branches were opened in Kathmandu and Delhi. While the Kathmandu Reception Centre is supported jointly by the UNHCR and CTA, the centres in Dharamsala and Delhi are funded by the CTA with a support from the United States' Government. The rehabilitation of the children below the age of 18 and monks and nuns is comparatively easier as the children can be sent to existing boarding schools and the monks and nuns to the monasteries and nunneries in India. The rehabilitation of adult refugees is problematic. This is because most of the adult new refugees are unskilled and most of the existing settlements are overcrowded. As such, the CTA opened a special Adult Education Centre in the outskirts of Dharamsala in 1993 to provide vocational and language training to the new refugees.

APPRAISAL AND REDESIGNING OF THE REHABILITATION PLAN

Over the years, the rehabilitation scene had changed considerably. The landholdings in the settlements have shrunken with the natural growth in the population and a new generation of educated and skilled population has thrown new opportunities and challenges. This requires new indepth appraisal of the on-going rehabilitation work in order to chart a new course for the future. A senior Indian official said that the problem of Tibetan refugees was basically a problem of three generations: the generation who came as refugees, the generation born in India and the generation that was still young. This posed peculiar problems of rehabilitation and reconstruction. He said, "While the problems are known, sometimes the answers are difficult to find. Or even if we know the answers, sometimes implementation is difficult." In view of this situation, a joint (New Delhi and Dharamsala) high-level workshop (or conference) was organized in New Delhi in October 1986 to review the situation and restructure the scheme. At the conference, the then Indian Home Minister, Mr. Buta Singh, emphasized the need for more diversification to target the needs of the second generation and timely execution of the rehabilitation programs both by the respective Indian State Governments and the Tibetans. An Indian expert on small-scale industries, said, "The present generation of Tibetan refugees having become aware that their traditional vacations are no longer going to help them in the long run, small-scale industries is an area of great potential which is yet to be introduced to this group of beneficiaries."

The conference, which was participated by the officials of the concerned Ministries of the Government of India, State Governments, and Tibetan officials, discussed in details various issues ranging from water supply and roads in the settlement to working capital for the cooperative societies, rehabilitation of additional refugees, education for children, acquisition of immovable property by Tibetan refugees, etc.

The Chief of the World Bank Mission, who also participated in the conference, emphasized the importance of maintaining cohesiveness of the community with physical proximity as this, he said, would help the Tibetan refugees to preserve their culture and identity in exile.

The UNHCR Representative assured that his agency would continue to provide "legal protection" to the Tibetan refugees, though "material assistance" ended in 1975 as the Government of India and UNHCR considered that the Tibetan refugees in India were self-sufficient economically.

In view of the above difficulties and challenges, efforts have been made since then to redesign and diversity the rehabilitation scheme.

The Planning Council was instituted in 1988 to improve the use of CTA's human, physical and financial resources in the development of the refugee community. The Council was instrumental in establishing the Handicraft Development Board and the Tibetan Computer Resource Centres, and introducing the concept of micro-enterprises and micro-credit in the community.

Efforts were made to improve communication and coordination among various sectors of the CTA, both in Dharamsala and peripheral settlements.

Recently, the Planning Council has conducted a detailed demographic survey of the Tibetan refugee community and random sampling survey of the socio-economic status of the community. These studies will definitely be of great use for the proper planning of the development initiatives in the community.

ARMED CONFLICT, REFUGEES AND THE ENVIRONMENT

Along with the destruction of lives and livelihoods, war can also destroy croplands, forests, water systems, and other natural resources. Clean air and soils were casualties of the 1990–91 Gulf War after being polluted when Iraqis intentionally ignited hundreds of oil wells. Marine and coastal life was damaged too; spills of 6–8 million barrels of oil into the Persian Gulf and Arabian Sea killed 15,000–30,000 sea birds and contaminated mangroves and coral reefs (UNEP 2002:14, 204, 292; Omar et al. 2000:317).

When Serbian forces systematically destroyed villages and towns in the 1999 Kosovo conflict, they also destroyed clean drinking water supplies and waste systems (UNEP and UNCHS 1999:5). And though decades have passed since U.S. forces cleared 325,000 hectares in the Viet Nam War by spraying the defoliant Agent Orange, biodiversity losses are still very much in evidence. Areas once covered by forests and mangroves now support just low-density grasslands and mudflats (McNeely 2000:362).

The toll on environmental governance is just as significant. War often destroys or weakens the institutions that make inclusive and informed decisions about the environment possible. The political and social turmoil that accompanies conflict can short-circuit systematic processes of environmental management.

War creates refugees, leaves government and environmental agencies handicapped or destroyed, and substitutes short-term survival for longer-term environmental considerations. This means that ecosystems continue to suffer even after the fighting has stopped.

War or "armed conflict" is a governance problem for a distressingly large number of people, ecosystems, and institutions. Between 1990 and 2000, 118 armed conflicts worldwide claimed approximately 6 million lives (Smith 2001:1). People and the environment suffered the consequences for years after the wars ended. In 1999, more than two thirds of the ongoing conflicts had lasted for more than 5 years, and almost one third had lasted for more than 20 years (Smith 2001:3).

Most current wars are fought within national borders, not between nations, but the effects often spill over to neighboring countries (CAII 1997; SIPRI 2002). Resource wealth is usually a factor in the violence, with competition for valuable resources like gold, diamonds, and timber driving the conflict. By one estimate, one quarter of the roughly 50 wars and armed conflicts active in 2001 were triggered, exacerbated, or financed by legal or illegal resource exploitation (Renner 2002:6).

Disrupted Governments

During and after conflict, governments generally focus on meeting immediate human needs—food, shelter, and safety for citizens and displaced populations. Protection of the environment and sustainable resource management are inevitably relegated to lower priorities.

Food shortages, disease, weakened health care systems, fragmented social networks, the destruction of people's livelihoods, and refugees who must be returned to their own homeland all take precedence over environmental concerns.

Even after conflict ends, well-informed environmental decisions are unlikely in the face of economic collapse, the need to rebuild infrastructure, and the disruption of commerce at the local, national, and international levels—common outcomes of armed conflict. War economies and destabilized governments perpetuate an ongoing cycle of violence and resource exploitation. Land and natural resources may be used as bargaining chips to gain allies during strife, in negotiations to end conflict, or as postwar paybacks to those who helped win the conflict. Little value may be accorded to intact ecosystems or ecosystem services in the process.

In times of conflict, governments and warring factions need money to buy arms and supplies; high-value resources such as ivory and diamonds can readily satisfy that demand. This dynamic has worked to the detriment of elephant populations in strife-torn countries such as Sudan, Chad, and the Central African Republic. It has also driven forest liquidation in Liberia and Sierra Leone. After the conflict ends, governments need to kick-start the economy and rebuild key sectors, and one of the quickest ways is to mine natural resources.

Armed conflict can wreak havoc on government conservation efforts, especially in protected areas. For example, during the Ethiopian-Eritrean war, parks and reserves lacked funds for staff, infrastructure, research, and management training. In countries where nature tourism provides a major source of income for biodiversity protection, that source quickly evaporates when conflict begins.

In Rwanda, income generated by tourists—many of whom come to see mountain gorillas—totaled about $4–6 million annually; this in turn funded conservation projects in parks and forest reserves. However, escalating conflict in the 1990s, and the 1994 genocide caused tourist numbers to plunge; they still have not fully recovered.

War often leads to the breakdown of law and order, leaving protected areas and species vulnerable to exploitation. During Sierra Leone's civil war in the 1990s, regional forestry officers, foresters, rangers, and guards went unpaid for long periods, while illegal mining and logging—and massive deforestation—occurred in forest reserves.

And while the Ethiopian-Eritrean war raged, game hunting by the military in protected areas continued.

In the Central African Republic, hunting and poaching in war-torn provinces reduced the country's elephant numbers by 90 percent to just 5,000 and led to the disappearance of the rhinoceros. And in Cambodia, the Khmer Rouge's trade in timber brought $10–20 million a month in funds for its civil war effort.

Even after wars end, weakened political institutions may not have the authority, ability, or funds to effectively manage their country's natural resources (Orr 2002:139). Some reconstruction efforts may include environmental projects, but they are not likely to be a priority.

Environmental ministries often lack the capacity to address environmental problems in any systematic way. The postwar turmoil can mean fragmented government ministries and new staff unaccustomed to working together or with other institutions. Years after the end of conflict in Bosnia and Herzegovina, environmental groups noted that new environmental legislation was forthcoming, but doubted the fledgling government's ability to implement and enforce it (REC 1997:35).

Local governments may be equally shattered, making it difficult to decentralize the management of natural resources effectively. Two decades of conflict in Afghanistan left local community decision-making bodies without the information, infrastructure, money, or human capacity to cope with demands on the environment (UNEP 2003:95).

3

The Challenges of Utilitarianism and Relativism

Human rights are usually said to be inalienable and universal, and some even believe that they should are absolute. Such attributes are necessary in order for human rights to protect all humans at all times. A prime motivation for rights in general is to ensure that no-one is subject to unbridled calculations of utility, so that a minority do not suffer in order that a great number enjoy some benefit. If anything is to stand in the way of governments or societies sacrificing individual or minority interests in favour of the collective, it is the bulwark of human rights. Similarly, human rights are argued to be universal and apply across political, religious, and cultural divides.

It is tempting in a liberal society such as Canada's to view human rights as both universal and inalienable. After all, so much of our political debate is built upon these suppositions that we take their reach for granted. However, these qualities of human rights may not stand up under the light of probing scrutiny. Human rights are particularly vulnerable to challenges from both utilitarianism and cultural relativism. These challenges relate to the nature of human *rights*, the choice of benefits that are said to be a matter of human rights, as well as the delivery of these benefits. Further problems emerge when one moves from the abstract right of an individual, to trying to assess the specific benefits any one individual is entitled to in relation to all others trying to exercise the same particular right, but the situation becomes even more complex when the issue involves balancing competing rights or balancing the good of individuals against the good of their community.

At one level rights are those claims which protect individuals from being subjected to calculations of pure utility. The promotion of the greatest happiness for the greatest number cannot justify some violation of an individual's welfare, if that individual has a right to the benefit in question. The most basic utilitarian critique of human rights lies in the assertion that resources are scarce in any society, and especially limited in some. This scarcity inevitably leads to utilitarian calculations to allocate those resources in a way that will maximize the greatest good. In the end, it is argued, all the benefits listed as human rights, even life itself, are subject to the promotion of the greatest good within a society. As such an individual's benefits claimed as a human right may be compromised, diluted, or even completely denied in specific situations where that right has to be weighed against the claim of another individual or of society as a whole. This critique is not necessarily normative, in the sense that this should be the case, but may also stem from the observation that this is how societies do and will function.

The utilitarian critique raises the question whether human rights are either absolute or inalienable. By inalienable, I mean that individuals cannot surrender control over their right to another's discretionary authority. The ultimate authority to make the most important choices with respect to exercising that right cannot rest with someone else - either the state, another individual, or some entity - but must be able to be reclaimed and exercised by the individual whose right is at stake. By absolute, I mean that the right in question cannot be totally denied. This is best seen in rights that pose dichotomous choices, (such as the right to life's "do you die or live?"), where the benefit is either provided completely or denied completely. It is difficult to argue that a right is absolute if the benefits it imparts can be enjoyed by degree - liberty is a classic example of a right that permits relative increases and decreases in its possession and exercise. Analysis becomes problematic since most rights are arguably entitlements to benefits that are exercised by increments. Thus, it becomes impossible to assert that all human rights are absolute. Nevertheless, one can suggest that at least one right is absolute, or at least should be if human rights are to have any substantive meaning taken collectively. The right to life is one such example, for no other human right can be relevant if life can be taken from an individual; the possession or enjoyment of all other human rights hinge on an individual being alive. Various examples illustrate the utilitarian foundation we eventually land against, but perhaps the most basic right, that to life, raises dilemmas for human rights theory if it cannot be shown to be absolute.

A starting assumption for a right to life that is absolute lies in arguing that innocent lives must be protected if human life has any value to be protected through human rights. Indeed, Alan Gewirth has argued that there must be at least one absolute right: "all innocent persons have an absolute right not to be made the intended victims of a homicidal project". If human rights cannot protect an innocent life from utilitarian calculations then one must question the force of these 'rights'. Gewirth portrays his argument with the example of an innocent mother held hostage by terrorists, who tell her son that they will detonate a nuclear explosion in a city if he does not kill his mother. According to Gewirth, the mother still has her right to life which the son must not violate. The son's duty and moral culpability lies solely in his own direct actions. The principle of intervening action means that the terrorists would be solely responsible for any deaths from their threatened explosion, since the son cannot be completely certain that the terrorists would carry out their threat. For Gewirth, his example of an absolute right stands the test. The son must not weigh the life of his mother against the lives of the city's population, because the other lives are not his responsibility. Gewirth's example, however, does not provide a scenario that fully tests the right of an innocent to life.

Utilitarian calculations on taking or sparing lives seem unavoidable in other situations. There is the classic case of a runaway trolley that can only be steered on two paths, one of which will run over one person and the other will run over five others. In that instance, the trolley driver would aim for the single individual. But this case is a highly unsatisfactory example, since the driver has no choice but to kill someone and would try to spare as many lives as possible. A more germane illustration is found in a SWAT team's arrival on a scene where a gunman is holding an innocent hostage as a shield with one arm while shooting into a crowd with the other. Should the police fire immediately to stop the gunman's killings, even if the hostage would likely be shot at the same time? Or, should the police allow the gunman to continue shooting while they manoeuvre to a vantage point where they can shoot the gunman without harm to the hostage? In this instance, the principle of intervening action invoked by Gewirth would mean that the police are not responsible for the deaths caused by the gunman.

Their direct duty is not to kill an innocent person themselves. They have a choice to kill the gunman and the hostage, or to wait and kill just the gunman. The police may even have the choice to simply wait until the gunman runs out of bullets and then tackle him without killing anyone themselves. Because they have the choice, they should not shoot the hostage just to stop the gunman killing others. However, many people simply would not agree with this approach. It may well be tragic, but justified nevertheless, for the police to shoot the hostage and gunman immediately rather than letting even more people be killed by the gunman. In this scenario, a utilitarian calculation to save several lives would outweigh the one innocent life. Thus, not even an innocent person's right to life appears absolute.

Whether there is an inalienable right to life, safe from the utilitarian needs of the state, is tested most sorely in times of war; but it is also as germane in times of peace. Considerable debate rages over the conscription of citizens to defend the state or pursue the state's interests abroad, but the right to life can be just as endangered for those citizens who voluntarily join the state's police, miliary, fire departments, and coast guard, and who are subject to superiors' orders that might lead to their death. Conscription raises the question whether the state can take control of its citizens lives and send them to their deaths. Voluntary service in the public safety and security services raise the issue whether individuals can contract away control over their lives for the duration of that service. In both instances, the issue is essentially whether the right to life is inalienable and cannot be given up to another's control.

Both Canada and the United States have had conscription during the twentieth century that pose a human rights challenge. Conscription was resorted to in both World Wars in Canada, although Canadian territory never sustained any direct attack beyond a few shells around a West Coast lighthouse. Largely for electoral reasons, the conscripts were mostly kept from the front lines, which were reserved for the volunteer enlistees. Nevertheless, the lives of the conscripts were at the disposal of the state and inevitably some died on duty, if even from traffic accidents. During the Vietnam War, many American men were drafted into a war effort thousands of miles from their homes; many thousands of these men went to their deaths. A ready objection to both the Canadian and American examples of conscription is that the conscripts were sent to fight a foreign war. But, does it matter that these examples of conscripts went off to wage war on a different continent? Would the state's control of these conscripted lives be any more justifiable if there was a direct and significant attack upon these peoples' homeland?

At issue is whether the state can claim command over the lives of its inhabitants and sacrifice them in the interests of the state. It would seem the state cannot, if there is an absolute human right that protects an individual's life. One might say there is a better reason for conscription if a state suffers direct attack, on the grounds of a collective self-defence. The state's right to sacrifice conscripted lives could arise in two ways, either because the state acts to exercise collectively the rights of its citizens to defend themselves, or because the state has rights greater even than its citizens due to its duty to defend those citizens. The first instance depends upon the right of self-defence of individuals. But, few if any would argue that there is a duty on an individual to defend their own persons. They may have the right to do so, in the sense of a privilege or immunity that can exonerate them from criminal responsibility for the harm they inflict in an act of self-defence. But that privilege does not impose a duty to defend oneself, and certainly not to the point of death. In the Hohfeld scheme of 'rights,' the right to self-defence is an immunity and not a claim-right with a correlative duty. Another dimension of self-defence lies in the duty-based moral rule that one should go to the aid of someone in distress. Could the state claim to enforce a duty of its citizens to defend others? But here, philosophers have not been able to agree whether any purported duty to rescue or defend others is a rule of internal morality or a rule of critical morality. In other words, do I have a duty to rescue or defend others only because I believe so, or because others believe I do? Whether it is an internal or external moral duty, there would be real disagreement over whether that duty extends to sacrificing one's life for another. Some religious individuals would believe it does; for example, Christians are told that there is no greater love than to lay down their lives for another. Many others, however, would disagree. In any event, such a purported rule is a *duty* upon individuals that is not correlative to some *right* that others can claim. Thus, it does not appear that the state can claim control of conscripted lives based on some analogy of the state as a corporeal aggregation of the rights of the individuals in that state. The state cannot claim with this logic rights that an individual does not possess.

Another justification for the state's ability to conscript citizens and sacrifice their lives might come from an argument that the state has some special rights that even individuals do not possess. And one of these is the right to command others to put their lives on the line in defence of the state - some might even extend that beyond the defence of the state to the pursuit of the state's interests. Particularly in the case of the defence of the state's very existence, one can argue that all the citizens will benefit if some are subject to conscription in a war that will preserve everyone's prosperity, way of life, or perhaps their lives. However, this line of reasoning is a complete negation of the prime justification for human rights - to protect individuals from being sacrificed for the benefit of the collective community (let alone the interests of their political leaders). There would be no human rights as they are generally viewed, if the state can claim special rights that are apart from and supersede the human rights of individuals. This justification for conscription is utilitarian to the core - the greatest good is served by requiring some individuals to submit to the state's control and sacrifice their lives.

It seems the only way to defend state conscription is if human rights are not viewed as inalienable. Thus, individuals can in certain circumstances either give up or lose their rights. They might give up their rights through a deliberate, voluntary gesture. Or, they may lose their rights because human rights might be alternatively viewed as part of a social contract that includes the possibility of losing rights - either through misbehaviour or through some greater power acceded to the their in the interests of protecting society as a whole. These different possibilities reflect different bases for human rights - either they inhere in humankind, or they are part of a contractarian foundation for civil society. If contractarian, human rights might be alienable or not, as the social contract of a society may or may not require.

There may be firmer ground for state control over citizens lives if human rights are really part of the social contract of citizens with their society. Human rights are best viewed as part of a social contract if they seen as logically required in any stable society - otherwise, they become civil rights not human rights. It may be argued that any civil (human-rights respecting) state might call on citizens to lay down their lives if the defence of civil society is at stake; human rights could logically be sacrificed in order for a society that human respects human rights to survive. Citizens owe a duty to defend the state because the state normally exists to defend and promote the rights and welfare of its citizens. Support for this position is found in Rousseau's view of the social contract, where individuals relinquish their rights to a state so that they can live under greater peace through the state's protection. But that peace can only come if the state is defended:

Their very lives, which they have pledged to the state, are always protected by it; and even when they risk their lives to defend the state, what more are they doing but giving back what they received from the state?...The purpose of the social treaty is the preservation of the contracting parties....Whoever wishes to preserve his own life at the expense of others must give his life for them when it is necessary. Now as no man is judge any longer of the danger to which the law requires him to expose himself, and when the prince says to him: 'It is expedient for the state that you should die', then he should die, because it is only on such terms as he has lived in security as long as he has and also because his life is no longer the bounty of nature but a gift he has received from the state.

With this contractarian a view of society, the state can call up its citizens and put their lives in jeopardy. The position that refuses to accept state conscription might spell the end of human rights, since it implies that human rights are so important that they must be respected right up until an authoritarian invader conquers the state and ends all human rights.

However, so much depends on the nature of the social contract. For example, a justification for conscription based on Rousseau's view of the social contract can be contrasted with Locke's, who held that the whole point of the social contract was to preserve the citizens and their property. If the state acted in a way to destroy their lives or property, then the citizens were justified in rebellion. Locke also believed that the powers of the state under the social contract could be no more than the powers of the individuals who joined together to form the state. As such, there are inherent limitations upon the power of the state:

First, it is not, nor can possibly be, absolutely arbitrary over the lives and fortunes of the people; for it being but the joint power of every member of the society given up to that person or assembly that is legislator, it can be no more than those persons had in the state of nature before they entered into society and gave up to the community; for nobody can transfer to another more power than he has in himself, and nobody has absolute arbitrary power over himself or any other, to destroy his own life or to take away the life or property of another. Social contract foundations for society unfortunately prove to be fertile grounds for disagreement. One can choose between two sides, either Locke's view of the a social contract and argue that extend the powers of the state may not extend to conscripting citizens and endangering their lives or Rousseau's view that the state can command its citizens to dies in its defence. Hobbes falls somewhere between Locke and Rousseau with a context-specific approach to what the state could demand of its citizens. He argued that the contract which founded the state could not generally endanger the lives of its citizens: "No man is bound by the words themselves, either to kill himself, or any other man." In some circumstances a man could be excused from responding to the state's call to arms, either by sending someone in his place or because of his 'timorous nature'. However, no-one can refuse a call to defend the very existence of the state: "When therefore our refusall to obey frustrates the End for which the Soveraignty was ordained; then there is no Liberty to refuse....And when the Defence of the Common-wealth, requireth at once the help of all that are able to bear Arms, every one is obliged." Thus in the divergence of views among Hobbes, Locke, and Rousseau we see very different approaches to the reach of the social contract in obliging citizens to submit to conscription.

Furthermore, not only are the terms of a purported social contract in doubt, but the manner in which an individual comes to be bound by the social contract is open to profoundly different theories. For Rousseau, only those individuals present at the state's formation get the opportunity to agree explicitly to the social contract: "After the state is instituted, residence implies consent: to inhabit the territory is to submit to the sovereign." Locke, on the other hand, believed that the social contract must be entered into specifically by each individual. Neither those born in a state's territory nor foreigners living in the society are automatically citizens. "But submitting to the laws of any country, living quietly and enjoying the privileges and protection under them, makes not a man a member of that society... Nothing can make a man so but his actually entering into it by positive engagement and express promise and contract." Not even a person born into a society would be a subject of the social contract until he or she reaches maturity and covenants for themselves.

However, another view of contractarian rights could hold that the individual's autonomy is so great that any individual may decide to do as they will with their rights. They may contract with the state in ways that threatens or extinguishes their rights, and even their lives. But can they contract to expose themselves to dangers or situations they never envisioned at the time of the contract? This is particularly relevant to the teenagers who enlist in the infantry with little idea of the twists that world or domestic politics will take during their 20 years of service. If teenagers can contract away their rights, is the sanctity of the contract so great that they cannot revoke it when faced with losing their lives in horrors they never imagined?

One can argue that when people volunteer for military service, even in peace time, that they voluntarily give up their right to life; they realize they agree that they can be ordered into situations that may will cost them their lives. This is an important consideration for countries who send their troops abroad to serve in international peace-keeping missions. Members of the military knew when they enlisted that they could be ordered into perilous circumstances that do not involve the direct protection of their own state. However, this justification depends upon the assumption that human rights are alienable. Some rights may be, but there may well be rights that mainly make sense only if they are inalienable, because their forfeiture can mean that they can never be exercised again. We may voluntarily set aside our rights for a temporary period, but not to give them up completely. The best example of this is our right to liberty. We frequently set this right aside for short periods; for example our right to express ourselves freely, or even to move about is sacrificed by most of us for the duration of a movie we see in a theatre. Quiet whispering is acceptable and so is getting up to go the washroom or buy more popcorn. But we usually give up the right to talk loudly or to get up and dance around. Indeed, the proprietor is usually in possession of these rights and may eject us if we try to exercise them. But such temporary surrendering of our rights is quite different from completely transferring them to someone else. The best example is found in surrendering our liberty and becoming slaves. Most people view voluntary servitude as being just as unacceptable as forced slavery. We cannot agree to become slaves, because our liberty is inalienable. For this reason debt bondage is equated with classic slavery and battled against by human rights activists. The same logic preventing us from voluntarily giving up our liberty to become slaves applies to our voluntarily giving up the right to our lives. If we are ordered into war zones by the state, there is a good chance that we will not live to exercise once again our independent control over our lives. The relinquishment of our right to life could become final and, for that reason, is unacceptable under a human rights regime that views human rights as inherent in our humanness. This problem of inalienability even pervades civil states at a time of peace. Police officers, fire fighters, and coast guard personnel are examples of groups of citizens whom the state expects to put their lives in danger for the greater good; military personnel assigned to international peace-keeping duties are another.

These different layers of discussion reveal that the justifiability of the state's control over its citizens depends a great deal upon one's view of the nature of human rights. Are they inherent and inalienable? Or is human society contractarian at best and individuals may either be required to give their lives for the state, or at least offer them up voluntarily? On the other hand, there may be a consequentialist concern with limits on the state's ability to defend the community. If the state cannot ultimately order its citizens to defend it, the values and benefits enshrined in human rights may be lost for generations. The consequence is that human rights may be a fragile ideal, easily lost to the next Hitler. In this view, the state has to have control over the lives of its citizens or many would face greater perils. A consequentialist view of contractarian revocation of rights would defend the state's ability to order its military to face dangers they had not bargained on in joining the forces. An inability to enforce discipline and the ability of soldiers to pick and choose which orders to follow would result in an almost indefensible state. Thus, the greater good requires that the state can order its military about as it sees fit. On the other hand, a different consequentialist concern may be that the state slides into authoritarianism in the fight to defend itself or that citizens are bound to respect even an authoritarian regime if rights are simply part of a social contract that gives a state the right to defend itself. These different alternatives underline the importance of the *choices* that must be made about human rights, and many of these dilemmas are difficult to resolve without concern for the greater good.

There is the very strong possibility that in fact the choices are ultimately settled by utilitarian calculations that balance up the needs of a stable society with individual autonomy. Certainly there cannot be a complex society without some individuals who put their lives in danger for others, be they police or firefighters. Many would also say that one cannot survive without some means to defend the state from invasion or authoritarian revolution, and one needs a military force that can be ordered into life-threatening situations. Just how many people are put into what dangers, and what ability they have to refuse the dangers may end up as pure utilitarian calculations. The greater good may be served by the state being able to issue orders that endanger and kill some of its citizens, in order to save the lives or freedoms of the rest of the population.

HUMAN RIGHTS AND UTILITARIANISM

Utilitarian challenges to the enjoyment of human rights need not occur only in such extraordinary circumstances. Imprisonment may be justified because there is thought to be a greater good for society that an individual be completely denied their freedom of movement and locked away. Utilitarian calculations may also resolve disputes that arise with conflicts between different rights or the enjoyment of the same right by different individuals. The decision faced by any government to balance the needs of health care, education, welfare payments, and the justice system leads to tough choices about the relative proportion of the state's budget that should be dedicated to each social program.

The distribution of state resources among these services will in the end depend on the government's perception of the greatest good provided for that society. Also, even within one area of spending the government will have to decide on distributing the benefits in a particular way. For instance, there may be a need to balance expensive hospital equipment, such as CAT scanners, against paying for nurses and hospital beds for patients undergoing general surgery. In the education system, governments need to balance the amount spent on primary, secondary, vocational, and higher learning. Different societies distribute their resources according their vision of the greatest good arising from the particular needs of that society.

One can respond in various ways to these challenges that utilitarianism pose to human rights. A simple way would be just to assert human rights provide a guide to how societies must try and re-order their priorities. Human rights are needed precisely because utilitarian calculations are widely made, often at the expense of some minority interest.

This response is motivated as much by an aversion to the perceived consequences of utilitarianism as by a commitment to human rights. Indeed, one's commitment to human rights may be motivated by a fear for a world where utility guides public policy. As R.G. Frey wrote, with classical utilitarianism "...there is no person who in principle is beyond the scope of utilitarian sacrifice". If utilitarian decisions are dedicated to promoting the greatest happiness for the greatest number, then it is conceivable that individuals or minority groups may pay a heavy price for the happiness of others in their community. Indeed, a time-worn objection to utilitarianism is that slavery could be justified in a society, if it produced sufficient good for the master class. Another objection is based on the view that individuals are not important and only have value in their contribution to the aggregate happiness of all members of the society taken together. However, these criticisms are based on an unbridled form of utilitarianism that has rarely been advocated since Bentham first posed the theory. Indeed, Mill's later development of utilitarianism contained inherent constraints; for example, he argued that such rights as freedom of expression were essential to determining utility.

Much of the objections aimed at utilitarianism are fostered by an obliviousness to different types of utilitarianism. While each author who writes on the subject seems to develop his or her own variety, two basic strands may be identified. The classic model is a kind of *act* utilitarianism that focuses on the utility produced by each separate action. Whatever act produces the greatest happiness is the alternative that should be followed. The other approach is *rule* utilitarianism, which essentially accepts that the greatest good is promoted in the long run by observing certain rules of behaviour even if a particular observance of a rule does not directly result in happiness. For example, it may be agreed that society is generally much better if people are honest and open about their mistakes and that they admit them to those who may have been harmed. In a given situation, I may have knowingly said unkind things about some friends that later led to their losing a business contract. If I confess to them that I was responsible, they would likely be very hurt and even end our friendship, but without regaining their lost contract. So, a confession in these circumstances leads to harm rather than happiness. While act utilitarianism would lead to a decision not to tell, rule utilitarianism would require me to confess despite the harm because society in general is better off if wrongs are admitted.

Some theorists have pursued rule utilitarianism as a way to reconcile human rights and utilitarianism. In this light, human rights become values that society believes must be consistently respected. Overall happiness is advanced for any given society if human rights are accepted as rules that structure policy-making and behaviour. Michael Freeden has argued that a constrained utilitarianism is perfectly compatible with human rights. Another avenue opened by Richard Brandt involves adapting the notion of 'rights' to utilitarian calculations. Brandt appears especially sceptical of the absolute nature of rights. He suggests that the nature of the obligation flowing from claim-rights is not absolute, but rather "not over-rideable by marginal or even substantial but only by extreme demands of welfare". With this view, human rights would normally be respected but may be set aside if other extremely important demands arise. For instance, a real threat of invasion would justify restricting political rights and diverting resources from social programs to national defence.

There are substantive difficulties, however, with these attempts to accommodate human rights and utilitarianism. James Fishkin has objected that a fundamental obstacle arises with the identification of the benefits to be protected as human rights. He argues that utility may underlie any attempt to select one set of human rights values over another. Furthermore, there can be little guarantee to safeguard against one of utilitarianism's perceived weaknesses: the benefits protected by rights may be distributed unevenly among the population in order to maximize society's collective gain.

Perhaps the most telling critique of attempts to reconcile utilitarianism with human rights is that the solutions proposed may end up leading not to universal human rights but to cultural relativism. Whether one refers to constrained utilitarianism or rule utilitarianism, the basic premise is that certain fundamental norms are said to frame utilitarian calculations, and these norms may be human rights. Utilitarianism is, in my view, a society-centred notion of policy choices - in another words the calculations for Canadians can only be made by Canadians, or for Fijians by Fijians. In order to accommodate universal human rights, one has to assert that each society must logically deduce that human rights benefits are as essential to their own. Thus, universalism might only be ascribed to human rights if each society recognizes their inherent value, or if they are necessary to the functioning of a complex human society. John Stuart Mill laid the groundwork for such a possibility, in his arguments that certain basic rights or liberties are essential for utilitarianism to function; freedom of expression and representative government, for example, are necessary for a society to debate and determine what the greatest happiness for the greatest number entails. However, this position is debateable, and one could argue that a benign sovereign may determine the greatest happiness without the trappings of representative democracy; traditional societies and even Marxist societies in the transitional socialist phase might be viewed in this light.

Moreover, there still remains the nagging question of what norms each society will end up adopting as the rules that must be considered. The very real possibility exists that societies will differ on just what benefits their citizens should enjoy in order to enhance the greatest happiness. Notions of equality will be expressed in very different benefits and circumstances for citizens of a non-theistic, liberal society than they will be in a traditional Islamic or Hindu society. In the end, rule or constrained utilitarianism may simply lead one down the path to cultural relativism, where each society determines for itself what basic norms must be protected and what sort of benefits may or may not be traded off in determining the greatest good for that society.

THE CHALLENGE OF CULTURAL RELATIVISM

Human rights face a serious challenge to their universality from cultural relativism. Since morality is inextricably linked to the general cultural values of a society, it is very difficult to argue that the moral standards arising from one society can be imposed on another. In its most extreme form, cultural relativism leads to the conclusion that each culture is equally valid and the ethical norms of any society are just as legitimate as those found in another society. Cultural relativism, therefore, poses a serious hurdle to global human rights standards: with the variety of political, religious, economic, and cultural values across the world, how can one set of 'human rights' bind all societies? The challenge raised by cultural relativism undermines the two dimensions of universalism: that all humans possess human rights, and that all humans enjoy roughly the same benefits from those rights.

There are several aspects of the cultural relativist challenges to all humans holding human rights. One fundamental question is whether all societies would agree on who is meant by 'humans' to which rights apply. However, different societies will draw different conclusions about who is 'human', and thus entitled to the protection of human rights. A number of cultures and religions have at times in their past viewed certain, or even all, outside groups as essentially sub-human barbarians who could never enjoy the status or rights of a member of that culture. Several societies have held that a member could lose whatever rights they held by some act of heresy or communication with the undesirables. But the issues of who is human or who can hold rights re-emerges in modern contexts in debates over the right to life of a foetus, deformed newborn, unwanted female baby, a murderer, comatose patient, and - in some societies - even those who renounce their religion. An important cultural difference emerges over the notion of claim-rights, which seems so central to most western conceptions of human rights. Some analysts have argued that this conception does not suit duty-based moral systems. Islam and Christianity are two of many religions whose morality is based upon duties imposed on the believers, without reference to the rights of others to be treated morally. The duty to treat others properly is owed to God and, it is argued, one cannot make a claim against the duty another owes to God. Islamic rulers are subject to dictates in the Koran about the treatment of their subjects, but these duties do not create any rights for the governed. In traditional Confucian philosophy, there are also duties placed on the rulers without rights being created. Without a claim-rights basis for human rights, there is little justification for any institution to adjudicate someone's claim that they have been denied a benefit that someone else was bound to respect. Another challenge to conceiving benefits as claim rights has come from Marxist theorists, some of whom have held that rights are a bourgeois creation needed as protection against the exploitive nature of the capitalist state, that should have no place in communist society.

Other Marxist theorists would allow for rights during the socialist transition to communism; however, in this view one cannot isolate the 'right' claimed by an individual from the inherent duty that also is held by that person. In this view, rights are not simply claims by one person against another's duty. The claimant also possesses a duty with respect to the benefit claimed; for example, one has not only a right to work, but a duty to do so as well. To pit an individual against the state in a rights claim is to divorce that citizen from their own duty, and - more importantly - to validate the individuals' antagonistic assertion that they have a better idea of their needs than the community leaders who are leading them towards a classless society. Depending on one's notion of the importance of a vanguard of the proletariat or of the value of any individual worker's judgment, one can either reconcile or deny that an individual's rights claim is compatible with Marxism or not.

The notion that human rights are needed to protect individuals from their society is not a concern felt in some cultures. A central challenge to human rights emerges from the perceived relationship of individuals to their society. Critics have argued that human rights are based on the liberal view that individuals remain distinct and severable from their society; indeed, the autonomy of the individual is one of the distinguishing characteristics of liberalism. But a number of other perspectives regard individuals as inseparable elements of their collective community. For example, traditional African cultures have been portrayed as placing more importance on the community than individuals, whose value stems from their contribution to the community as a whole. Even in Western cultures, there are many who do not share liberalism's emphasis on the worth of each individual. Classical conservatism, for example, stressed the manner in which individuals belong to an hierarchical, organic society in which each individual assumes importance in the way in which they can contribute to the collective society's welfare. The greater needs of the whole society may require a stratified society with quite different roles, duties, and benefits that would seem incompatible without many ideas of liberal equality; the conservative notion of equality may stress equality in the value of individuals rather than any equality of opportunity let alone condition.

An objection common to all these positions lies in the significance other perspectives place on the relationships of individuals to their society. Human rights have a completely different focus, with their emphasis on the benefits which individuals are entitled to claim against their community. But many cultures, religions, and political philosophies are not immediately embracing or even compatible with such importance being attached to the primacy of the benefits to be accorded by human rights to any one ordinary individual within their society.

Even if one can move beyond these fundamental disputes, cultural relativism persists as a thorny obstacle when it comes to deciding what particular benefits should be protected within the human rights rubric. For example, some of the most persistent problems that human rights advocates see is the lack of equality in societies around the world. Unfortunately, there are profoundly different ways in which "equality" may be conceptualized. Even with an agreement upon a particular form of equality, there still remains a failure of many cultures to accept liberalism's insistence that equality is essential.

Indeed, many beliefs centre on individuals occupying different stations in life and that involve particular responsibilities and privileges. Confucianism involves a hierarchically structured society. Islam distinguishes between believers and two classifications of non-believers.

Hinduism has held that people are born to different castes because of their spiritual progress in their past lives. And, the Roman Catholic church and some Protestant churches insist on male supremacy within their institutions and liturgy. There are many across the world who will ardently resist any imposition of full equality in all aspects of their society. With this in mind, there will be profound disagreements about whether there is a right to equality inherent in either human existence or dignity. There may be even more grounds for disagreement over which particular conceptualization of equality to pursue - equality of condition, situation, or opportunity and so on.

Even where there is agreement on a principle that should be embodied in a right, there can be widely differing views of the substantive benefits that must be provided in order to respect the principle. For instance, if there is a right to education, does this right involve compulsory, free education? If so, up to what level? And, who has control over what subjects are taught and by whom? Are adults entitled to continue an interrupted education whenever they want, and at whose expense? Since a number of religious groups believe that education should take place within a completely spiritual framework, must the state allow, or even fund, separate school systems? These questions will be answered quite differently depending on whether the society is, for example, industrialized or agrarian, Shiite or largely atheistic, harmoniously plural or rigidly homogenous.

Thus, cultural relativism challenges human rights in their conception as well as their operation. If human rights are to pose some global standard, then there must be some response to relativism. Jack Donnelly has drawn up a typology of a spectrum of relativism that usefully shows the range of the challenge to universal moral precepts such as human rights. With the extreme form, radical cultural relativism, each society's culture is the unique foundation for its morality. Human rights are seen to be just western liberal mores that some seek to impose on other cultures. Pollis and Schwab have argued this point to declare that human rights cannot have universal legitimacy. However, most writers reject the notion that each culture has absolute authority. Rhoda Howard argues that this perspective romanticizes the value of various cultures in a way that permits terrible abuses: "Cultural absolutism forgives cruelty on the grounds that acting in accordance with the customs of one's own group is a universal moral principle". With radical cultural relativism, other societies could not legitimately censor the domestic policies of Nazi Germany, apartheid in South Africa, or ethnic cleansing in the former Yugoslavia.

At the other end of the spectrum, radical universalism asserts that there are indeed universal moral rules that cannot be legitimately overridden by the cultural values of particular societies. From this perspective, human rights would be global standards from which no society could depart; all cultures would have to protect and provide the same benefits that human rights safeguard. Unfortunately, some human rights advocates stray, sometimes inadvertently, into this position; as Donnelly and Howard once wrote:

Internationally recognized human rights require a liberal regime. Other types of regimes, and the conceptions of human dignity on which they rest, may be defensible on other moral and political grounds, but they will not stand up to scrutiny under the standards of human rights.

While Donnelly and Howard may not wish to be characterized as universalists, this quotation illustrates a serious problem with global human rights standards. So much of the values enshrined in contemporary human rights documents are liberal in origin that they require a liberal regime to be realized fully. As some critics charge, human rights really may be an attempt to universalize liberalism as the ultimate standard to which all societies should conform. It is one thing to argue that liberalism is a preferred form of society, but it quite another to argue, even implicitly, by appeals to human rights that every human is inherently entitled to the benefits of a liberal society simply by being human.

Many commentators would probably prefer the middle ground between radical cultural relativism and universalism, but even this range has its problems. Donnelly identifies two intermediate categories, strong cultural relativism and weak cultural relativism. In the former, each society's culture provides the principal basis for its moral standards, while in the latter each culture is just an important source of its morals. In these categories, external mores play some role in the formation and criticism of any particular society's values. In strong cultural relativism, each culture forms the base of its mores and human rights would be "a check on potential excesses of relativism". With weak cultural relativism, however, external mores such as human rights would provide the main ethical framework, which local cultural values modify slightly.

While there are good arguments to reject both radical cultural relativism and universalism, it is rather difficult to construct a watertight argument that either strong or weak cultural relativism must be rejected in favour of the other. Donnelly tries to settle the issue by framing the discussion around the nature of the effect that culture would have. He says that cultural values could shape the substance, interpretation, and form of human rights. But Donnelly concludes that universal human rights should vary only in the form in which each culture would deliver the benefits at issue.

Unfortunately, this resolution depends on one's acceptance of the premise that there should be universal human rights in the first place, and that the benefits should be roughly similar across the world. Some human rights advocates, such as Donnelly, simply dismiss the strongest challenges of cultural relativism - to the existence of human rights in general or of particular rights - as being irrelevant; they argue that human rights evidently do exist and certain benefits must be provided to everyone. A number of other writers have approached the challenge of cultural relativism from another perspective, by searching for common ground among the world's cultures. Alison Renteln has also asserted that human rights can be accommodated within a variety of cultures if a wider view is taken of the nature of human rights. There are arguments that human rights norms can be quite reconcilable with Marxism, Islam, or traditional African, Chinese, and native American cultures. In some respects this avenue of inquiry becomes an empirical anthropological search for moral commonalities.

If one can find moral goods or benefits that are common, then these may form the basis for universal human rights. But there is a logical flaw in this approach, since a survey of moral practices is just as likely to turn up many values upheld in societies that would horrify human rights activists - infanticide, banishment, and capital punishment are but three examples. Rhoda Howard has validly noted that there have been so many cultures that have inflicted cruelty on their members or strangers that human rights cannot be simply established by saying that every culture has respected the essence of human rights in their own way. A substantial challenge to a search for moral universals through anthropological empiricism lies the argument long made by feminists- that some of the most enduring traits of organized human society have been patriarchy and the oppression of women. While there are many examples of matriarchal societies, they are few in comparison to patriarchies. If so many societies over the millennia have believed that men should rule and women follow, what view of equality would we be led to adopt from the anthropological evidence? I doubt it would be one that would bring comfort to most liberal human rights activists.

There is a more revealing side to this debate, in the weight to be given to cultural commonalities. If one were to search for common moral values that could ground human rights, it is doubtful that absolutely universal practices or values would emerge. The most one is likely to say is that the overwhelming majority of societies have held such-and-such a view. If a universal moral value is to built upon a preponderant, rather than universal, support, then moral rules risk simply becoming the domain of the majority to which minority cultural practices must conform. But if such a moral majoritarianism is to be admitted, then patriarchy must be adopted as the determinant ingredient of gender equality because of the shear weight of its practice.

Moreover, the end of such a debate over moral commonalities is more likely to lead to a list of shared moral goods or benefits that are valued in societies, rather than a complete rights paradigm. For example, traditional Christian and Islamic societies share many notions of benefits that individuals should be given, but they founded such benefits upon God-given duties and not upon the inherent entitlements of the person who would enjoy the benefit. The claim-rights conception of human rights would be difficult to ground in common moral or cultural practice. In large measure, attempts to reconcile contemporary human rights with different value systems may depend on accepting that the benefits involved need not necessarily be protected only as claim-rights; duty based moralities can provide similar substantive enjoyment of benefits as a rights-based approach. However, it is sometimes necessary to admit that certain benefits would not fit within a particular culture's list of human rights benefits; equality and freedom of religion can be especially problematic to reconcile in the same way in different societies.

Human rights face significant hurdles when one tries to apply them universally across cultural, religious and political divides. The full force of the problem emerges clearly when a particular cultural or religious group is said to flagrantly infringe a right embodied in widely-supported human rights documents, such as the Universal Declaration of Human Rights. The members of that culture claim that their practice is a revered moral value within their society and that the outside world has no moral authority to require a change to conform to the external norms. The problem is especially compounded when the infringement involves some issue that defines that culture or religion. For example, one of the most egregious sins for a Muslim is to renounce Islam and take up another religion; many Christians would also say that a Christian risks eternal damnation by turning his or her back on Christ. And yet the Universal Declaration of Human Rights specifically states in Article 18 that the right to freedom of religion includes the freedom to change one's religion or belief. So, does any religious community infringe human rights if it punishes or condemns apostasy? Those who believe there is only one, true religion cannot be easily persuaded that an adherent has the right to renounce that religion and adopt another faith; the apostate may lead others to renounce their one path to salvation. An yet, the UNDHR would suggest that as a matter of human dignity, every individual has the human right to change religions. As such they ought to be encouraged in their change of beliefs as an exercise of their fundamental rights - not condemned, punished, or prevented from converting others.

A fundamental dilemma for both universalists and cultural relativists lies in the tension between external and internal morality. On the dimension of interaction between different cultural groups, the relativist would argue that the external morality is of no concern to the members of any particular group; only the internal morality of that group is legitimate. The universalist would have to argue that internal morality of any one group is subordinate to the external morality of the whole collection of groups. In the human rights context, this means that a relativist would argue that international human rights codes cannot be imposed upon any particular society, since only that society's own moral value system has legitimate authority there. In contrast, the advocates of universal human rights would say that the practices of any particular society must be changed to conform to the moral standards of international human rights.

However, the interplay between internal and external morality can change dramatically when one moves the analysis from the dimension of global society versus an individual society to the dimension of any particular society versus an individual in that society. Whereas relativists have to uphold the primacy of internal morality in the global v. individual society dimension, they may have to accept the primacy of external morality when looking at conflict between a society and an individual member. If relativists argue that only each society may legitimately establish the moral rules for that society, then they cannot object to the legitimacy of a groups' cultural rule that demands conformity from the members of that society - such as a prohibition against apostasy on pain of death. On the other hand, universalists must champion the cause of external morality at the global level, but the nature of human rights means that they must allow the primacy of internal morality in many instances where an individual's internal morality clashes with the mores of her or his society.

UNIVERSAL HUMAN RIGHTS AND CULTURAL RELATIVISM

This situation sharpens a long-standing dilemma: How can universal human rights exist in a culturally diverse world? As the international community becomes increasingly integrated, how can cultural diversity and integrity be respected? Is a global culture inevitable? If so, is the world ready for it? How could a global culture emerge based on and guided by human dignity and tolerance? These are some of the issues, concerns and questions underlying the debate over universal human rights and cultural relativism.

Cultural relativism is the assertion that human values, far from being universal, vary a great deal according to different cultural perspectives. Some would apply this relativism to the promotion, protection, interpretation and application of human rights which could be interpreted differently within different cultural, ethnic and religious traditions. In other words, according to this view, human rights are culturally relative rather than universal. Taken to its extreme, this relativism would pose a dangerous threat to the effectiveness of international law and the international system of human rights that has been painstakingly contructed over the decades. If cultural tradition alone governs State compliance with international standards, then widespread disregard, abuse and violation of human rights would be given legitimacy.

Accordingly, the promotion and protection of human rights perceived as culturally relative would only be subject to State discretion, rather than international legal imperative. By rejecting or disregarding their legal obligation to promote and protect universal human rights, States advocating cultural relativism could raise their own cultural norms and particularities above international law and standards.

UNIVERSAL HUMAN RIGHTS AND INTERNATIONAL LAW

Largely through the ongoing work of the United Nations, the universality of human rights has been clearly established and recognized in international law. Human rights are emphasized among the purposes of the United Nations as proclaimed in its Charter, which states that human rights are "for all without distinction". Human rights are the natural-born rights for every human being, universally. They are not privileges.

The Charter further commits the United Nations and all Member States to action promoting "universal respect for, and observance of, human rights and fundamental freedoms". As the cornerstone of the International Bill of Rights, the Universal Declaration of Human Rights affirms consensus on a universal standard of human rights. In the recent issue of *A Global Agenda*, Charles Norchi points out that the Universal Declaration "represents a broader consensus on human dignity than does any single culture or tradition".

Universal human rights are further established by the two international covenants on human rights (International Covenant on Economic, Social and Cultural Rights, and International Covenant on Civil and Political Rights), and the other international standard-setting instruments which address numerous concerns, including genocide, slavery, torture, racial discrimination, discrimination against women, rights of the child, minorities and religious tolerance.

These achievements in human rights standard-setting span nearly five decades of work by the United Nations General Assembly and other parts of the United Nations system. As an assembly of nearly every State in the international community, the General Assembly is a uniquely representative body authorized to address and advance the protection and promotion of human rights. As such, it serves as an excellent indicator of international consensus on human rights. This consensus is embodied in the language of the Universal Declaration itself. The universal nature of human rights is literally written into the title of the Universal Declaration of Human Rights. Its Preamble proclaims the Declaration as a "common standard of achievement for all peoples and all nations".

This statement is echoed most recently in the Vienna Declaration and Programme of Action, which repeats the same language to reaffirm the status of the Universal Declaration as a "common standard" for everyone. Adopted in June 1993 by the United Nations World Conference on Human Rights in Austria, the Vienna Declaration continues to reinforce the universality of human rights, stating, "All human rights are universal, indivisible and interdependent and interrelated". This means that political, civil, cultural, economic and social human rights are to be seen in their entirety. One cannot pick and choose which rights to promote and protect. They are all of equal value and apply to everyone.

As if to settle the matter once and for all, the Vienna Declaration states in its first paragraph that "the universal nature" of all human rights and fundamental freedoms is "beyond question". The unquestionable universality of human rights is presented in the context of the reaffirmation of the obligation of States to promote and protect human rights. The legal obligation is reaffirmed for all States to promote "universal respect for, and observance and protection of, all human rights and fundamental freedoms for all". It is clearly stated that the obligation of States is to promote universal respect for, and observance of, human rights. Not selective, not relative, but universal respect, observance and protection.

Furthermore, the obligation is established for all States, in accordance with the Charter of the United Nations and other instruments of human rights and international law. No State is exempt from this obligation. All Member States of the United Nations have a legal obligation to promote and protect human rights, regardless of particular cultural perspectives. Universal human rights protection and promotion are asserted in the Vienna Declaration as the "first responsibility" of all Governments.

Everyone is entitled to human rights without discrimination of any kind. The non-discrimination principle is a fundamental rule of international law. This means that human rights are for all human beings, regardless of "race, colour, sex, language, religion, political or other opinion, national or social origin, property, birth or other status". Non-discrimination protects individuals and groups against the denial and violation of their human rights. To deny human rights on the grounds of cultural distinction is discriminatory. Human rights are intended for everyone, in every culture. Human rights are the birthright of every person. If a State dismisses universal human rights on the basis of cultural relativism, then rights would be denied to the persons living under that State's authority. The denial or abuse of human rights is wrong, regardless of the violator's culture.

HUMAN RIGHTS, CULTURAL INTEGRITY AND DIVERSITY

Universal human rights do not impose one cultural standard, rather one legal standard of minimum protection necessary for human dignity. As a legal standard adopted through the United Nations, universal human rights represent the hard-won consensus of the international community, not the cultural imperialism of any particular region or set of traditions.

Like most areas of international law, universal human rights are a modern achievement, new to all cultures. Human rights are neither representative of, nor oriented towards, one culture to the exclusion of others. Universal human rights reflect the dynamic, coordinated efforts of the international community to achieve and advance a common standard and international system of law to protect human dignity.

Inherent Flexibility

Out of this process, universal human rights emerge with sufficient flexibility to respect and protect cultural diversity and integrity. The flexibility of human rights to be relevant to diverse cultures is facilitated by the establishment of minimum standards and the incorporation of cultural rights.

The instruments establish minimum standards for economic, social, cultural, civil and political rights. Within this framework, States have maximum room for cultural variation without diluting or compromising the minimum standards of human rights established by law. These minimum standards are in fact quite high , requiring from the State a very high level of performance in the field of human rights. The Vienna Declaration provides explicit consideration for culture in human rights promotion and protection, stating that "the significance of national and regional particularities and various historical, cultural and religious backgrounds must be borne in mind". This is deliberately acknowledged in the context of the duty of States to promote and protect human rights regardless of their cultural systems. While its importance is recognized, cultural consideration in no way diminishes States' human rights obligations.

Most directly, human rights facilitate respect for and protection of cultural diversity and integrity, through the establishment of cultural rights embodied in instruments of human rights law. These include: the International Bill of

Rights; the Convention on the Rights of the Child; the International Convention on the Elimination of All Forms of Racial Discrimination; the Declaration on Race and Racial Prejudice; the Declaration on the Elimination of All Forms of Intolerance and of Discrimination Based on Religion or Belief; the Declaration on the Principles of International Cultural Cooperation; the Declaration on the Rights of Persons Belonging to National or Ethnic, Religious and Linguistic Minorities; the Declaration on the Right to Development; the International Convention on the Protection of the Rights of All Migrant Workers and Members of Their Families; and the ILO Convention No. 169 on the Rights of Indigenous and Tribal Peoples.

Human rights which relate to cultural diversity and integrity encompass a wide range of protections, including: the right to cultural participation; the right to enjoy the arts; conservation, development and diffusion of culture; protection of cultural heritage; freedom for creative activity; protection of persons belonging to ethnic, religious or linguistic minorities; freedom of assembly and association; the right to education; freedom of thought, conscience or religion; freedom of opinion and expression; and the principle of non-discrimination.

LINGUISTIC AND CULTURAL RIGHTS AND RESOURCES

Since 1992, under the Sami Language Act (516/91), Sami people have had the right to use the Sami language before authorities, orally and in writing, and to receive a reply in the same language. In practice, however, the exercise of these linguistic rights by the Sami is based on translation and interpretation, creating crucial obstacles to dealing with authorities. The level of linguistic rights afforded to the Sami population in the Sami Language Act should be re-examined to assess whether it reaches the level guaranteed to the Sami in the Constitution Act as regards the right to maintain and develop their own language and culture. The status of the Sami language in school laws should be examined in the same light.

In Inari the education and training centre for the Sami region already offers classes in the Finnish and Sami languages; the knowledge of both languages is included in the qualifications of newly recruited teachers. The purpose of the school is to preserve and develop the Sami culture and natural livelihoods. The Sami people are in a determining position in the administration of the training centre.

In 1996 the Sami Parliament adopted three Sami languages spoken in Finland as the languages which have an equal status in the Parliament: Inari Sami, Skolt Sami and Northern Sami. The implementation of the decision is being postponed until government funds are made available for it. In many places Semi people have already changed. languages. This particularly applies to Inari Sami and Skolt Sami. and in certain areas also lo Northern Sami. The revival, maintenance and development of the Sami languages require speedy measures and targeted resources.

Today. the practice of Sami culture is subsidized from the state budget. Decisions on the allocation of the funds are made by the Sami Parliament. The Sami Radio has a channel for daily broadcasts in the Sami language. Construction for a Sami Museum started in the autumn of 1996. Special measures are still needed to create such cultural services as TV broadcasts in the Sami language, especially programming for children, to support Sami arts, and to protect Sami handicrafts and the Sami cultural heritage.

4

Environmental Regulation and Human Rights

Our intention is certainly not to review the whole of environmental law, which has during the last two decades evolved into one of the most dynamic and expanding branches of international as well as national law, but to reconsider only some problems which are more interesting than many others from the standpoint of the human rights probl-matique. It should be stressed, however, that all regulation in this field impinges directly or indirectly on human rights. It may be action against desertification, attempts to lessen or stop acid rain, better control of food production, efforts to make human settlements more habitable, or any other activity in the vast field of the environment, but it always protects or improves some human right.

Although this development has been rapid it is still inadequate in many fields, in spite of the fact that basic principles have been developed. Serious efforts at regulation in a more concrete way (by international conventions, by institutional arrangement, and in other ways) must be undertaken, on a global, regional, sub regional, or bilateral basis. That means more concrete legal regulation at all levels, as well as additional work on elaboration of basic principles, and in the first place of the principle of "liability and compensation for the victims of pollution and other environmental damage," as formulated in the Declaration of the United Nations Conference on the Human Environment (Principle 22). Besides that, it should be kept in mind that states are responsible not only towards other states and the international community as a whole, but also towards their own citizens, who have the right to a healthy environment.

The enjoyment of all human rights is closely linked to the environmental issue. Not only rights to life and health in the first place, but also other social, economic, cultural, as well as political and civil rights, can be fully enjoyed only in a sound environment. And certainly, to go to an extreme, they cannot be enjoyed at all if the environment becomes impaired beyond a certain critical level. The whole of mankind could in such a case perish together with all its civilization, including human rights. The worse the environment becomes, the more impaired are human rights, and vice versa. That is the reason why there is the need for sustainable development and that means, in the first place, ecologically sound development of economies, science and technology, and all other fields. This is a sine qua non for both protection of the environment and further promotion of human rights. Besides the undeniable interdependence between the environmental issue and all human rights, a new human right - the right to an adequate environment - is emerging. This right, still not precisely formulated, appears in documents and in literature, in some cases as a collective and in other cases as an individual human right.

The Right to an Adequate Environment

The right to an adequate environment or, as it is termed in some texts, a satisfactory environment, is one of the so-called third-generation or solidarity rights. It can be found in international documents of both a declaratory and formally binding nature, as well as in domestic legislative and other acts of a number of countries, including some constitutions. The African Charter, for instance, proclaims that: "All peoples shall have the right to a general satisfactory environment favourable to their development." In the Declaration of the United Nations Conference on the Human Environment (Stockholm, 1972) it appears, however, also as an individual right. The Declaration states that: "Man has the fundamental right to freedom, equality and adequate conditions of life in an environment of a quality that permits a life of dignity and well-being and he bears a solemn responsibility to protect and improve the environment for present and future generations." It appears as an individual right also in the report of the World Commission on Environment and Development, which proposes, as one of the legal principles for environmental protection and sustainable development, that:

> *"All human beings have the fundamental right to an environment adequate for their health and well-being." Finally, it should be noted that elements of this right can be found in the Universal Declaration of Human Rights, as well as in both Covenants, although the environment as such is scarcely mentioned in the documents.*

It should be added that individuals, as well as groups, not only have the right to an adequate environment, but also the duty to protect and improve the environment. They have this responsibility not only towards other individuals or the community in which they live but towards mankind as a whole and even "future generations." It is a responsibility which certainly collides in many cases with the enjoyment of their other rights, be they common citizens or those who, as scientists, technicians, decision-makers or in any other way, are more closely linked to scientific or technological development, environmental protection, health protection, or other cognate fields.

Definitional Problems

Many questions arise in connection with this right, when it is formulated as a collective right, and some of these questions are similar to the questions concerning other human rights of the so-called third generation In the first place it is not easy to see "how individuals can assert it against states, and so how it can be satisfactorily classified as a human right." Besides that, we still do not have a precise definition of the right, whether as a collective right ("the right of all peoples," or of "people," or of all mankind, or of future generations) or as an individual right (right of man or of all human beings). Paul Sieghart believes that in the case of all "third-generation" rights, including the right to an adequate environment, "some formulation will have to be devised whereby each of them can be clearly seen to vest in individuals, to be exercisable by individuals, and to impose precise correlative duties on states so that it can then be interpreted, applied and enforced accordingly." It is certainly true that further work on the "third-generation" rights is needed if we wish to improve the whole system of international legal protection of human rights.

Developments in the field of the environment, which are faster than in many other fields, will without doubt contribute to the establishment of a balance, as well as more links, between the first and second "generations" of human rights, as well as clarify the relation between individual human rights and the "third-generation" or collective rights. It is necessary to treat the whole problmatique of human rights as a system in which all components are interrelated and play specific and functional roles. We stress the statement in the UN Declaration on the Right to Development that "equal attention and urgent consideration should be given to the implementation, promotion, and protection of civil, political, economic, social, and cultural rights and that, accordingly, the promotion of, respect for and enjoyment of certain human rights and fundamental freedoms cannot justify the denial of other human rights and fundamental freedoms." Further development of the right to adequate environment will certainly contribute to these efforts.

One of the most important measures in the field of human rights is, without doubt, further work on a more precise definition of specific human rights, including the right to an adequate environment. Each of them must be explicitly defined, especially in national jurisdictions. While some human rights are relatively easy to define, others are not. The specific right to privacy, to mention only one example, "had not been explicitly defined in many countries." Without setting precise definitions, efficient regulation by legal and other means is difficult, if not impossible, to imagine. It must be added that on the international level, when it comes to their definition, the rights which have newly emerged (like the right to adequate environment or the right to development) and emerging human rights deserve special attention. The work on definitions should be done not only by legal experts but by natural scientists, technicians, and other experts on a multidisciplinary basis. In many cases scientific and technical expertise is needed and various rules, such as ecological standards, are necessary as an addition to the more classical legal definitions and regulations.

The Collision of Rights

The right to an adequate environment, by its very nature, is one of the rights colliding with other rights, in a certain sense actually playing the role of a controlling mechanism for the enjoyment of other rights. In a way it sets the functional limits of all other human rights, especially if the environment is defined in a broader sense, as we believe it should be. In short, if any human activity impairs the environment beyond a certain limit, it should be regarded as an activity producing negative results and be forbidden or changed until it ceases to produce such detrimental results. On the other hand, if human activities do not impair, or if they improve the environment, they should be regarded as beneficial.

Institutional Machinery

Efforts should be continued, especially within the United Nations system, for developing a more efficient machinery for implementation of human rights, on a universal as well as on a regional, subregional, and bilateral basis. That means that the United Nations should be more oriented to practical action, even before any more serious UN reorganization is undertaken. This does not mean, however, "that the legislative process in the field of human rights should be halted, if not ended" at this phase, as some authors believe. It seems to us that even some classical human rights could be further developed and more clearly defined.

One of the reasons for this view is the necessity for better protection of human rights from the negative influence of the development of science and technology. Another reason, closely related to the first the impact of the environmental probl-matique. A third is its essentiality for the whole field of development. Besides, further development of the newly emerged human rights, for instance the rights to adequate environment and to development, certainly requires not only that they be more precisely defined but that they be legally regulated in a more concrete and efficient way. States should undertake a solemn obligation to take part individually and through international co-operation in promoting specific programmes and all other appropriate measures in the field of human rights, having in mind the use of science and technology for further protection and promotion of human rights, the environmental issue, and development in economic and other fields.

The Role of Non-binding Documents

International documents of a non-binding nature play an important role in all fields of international relations, and the human rights problematique is not an exception. In many cases they can regulate international relations in a specific field de facto, although they are not formally binding. Besides, they often lead to a higher level of regulation, i.e. international treaties, institutional arrangements, etc. The International Declaration on Human Rights was followed, for instance, by the Covenants, and similar developments occurred in other fields. From that point of view the proposed Declaration on Human Rights and Scientific and Technological Developments could be an important step towards better protection and promotion of human rights. It seems to us that such a declaration should contain, more or less, what has already been proposed by the group of experts convened by the United Nations in Geneva in September 1975, but that an additional effort should be made to link it more closely to development and the environment.

Naturally, one must always have in mind the other side of the coin More so-called soft law does not automatically mean more regulation and further progress in the field. In some cases, it leads to a proliferation of documents with very little or no importance at all, and it could even hinder the process of legal regulation. In other cases, however, including the fields of science, technology, the environment, and some other cognate domains, non-binding norms can be, if functionally well designed, very important and almost universally accepted and applied in such a way as to influence human rights beneficially. Good examples of such norms are formally non-binding ecological standards, which are accepted by all or most interested states and other subjects of international law because it is in their interest to do so. The sanction for those who do not apply the standard becomes "functional." That means that the mere fact of not applying them can cause impairment of the environment, loss of profit, health problems, or even loss of life, lessening of political prestige, etc. If adequately set, they could represent an optimal mode of behaviour. In that way, although formally non-binding, they contribute to the protection of the environment, having a directly or indirectly positive impact on the protection and promotion of human rights.

Military Technologies

Although this paper is not devoted to the military aspect of the problematique, a few words should be said. There is no doubt that the advancement of military technologies is one of the greatest dangers to all human rights, including the most basic right to life itself. The development of nuclear weapons, and other arms for mass destruction, as well as military technologies in general, poses a very serious threat to the environment and *ipso facto* to human rights, even if not used at all in war (radiation caused by nuclear tests, pollution produced by the military-industrial complex, impairment of the environment originating from military manoeuvres, etc). In the case of a large-scale use of such weapons, extinction of humankind as well as destruction of the environment, including all or most of life on planet earth, is possible. Production of various other kinds of sophisticated weaponry is in many cases meant more for domestic use (i.e. various kinds of oppression). Introduction of new technologies, be they new weapons or other police devices, used against criminals who by their illegal conduct are impairing the human rights of other members of society, is justified if based on legal procedures and obedience to democratic laws by police and other state services. It is, however, certainly not acceptable wherever such means are used for oppression, based on discrimination of any kind mentioned in Article 2 of both Covenants.

It would be ideal if we could achieve general and complete disarmament and use at least a part of the financial resources thus released "for comprehensive development, in particular that of the developing countries." The realities of our time are not in favour of such an ideal solution. General and complete disarmament is still not a thing of the near future, but let us hope that the efforts which are being made will bring at least partial success, more efficient international control and better relations among the two opposing military blocs, as well as the end of various local armed conflicts. Otherwise, especially where nuclear and other weapons for mass destruction are concerned, our future, including further action for protection and promotion of human rights, will not be very bright. It is interesting to mention, however, that the development of science and technology can help to a certain extent ever before total disarmament is attained.

For instance, the "technical progress over the past two decades has reduced the prospects of purely accidental war." The same progress, and especially the use of space technology for remote sensing of the earth, is already helping superpowers to control each other, and it will make possible, combined with inspections and other means, the control of any future disarmament plan. Finally, it is interesting to add that in some cases improvement of technology can, even in war, prevent unnecessary human losses and suffering. Joseph S. Nye Jr believes, for example, that the increase of accuracy with which weapons can be delivered "could to reduce reliance on battlefield nuclear weapons and on city-burning strategic weapons. "

Ultra-hazardous Activities

Another important issue covers all potentially very dangerous activities, usually called ultra-hazardous. They could cause (and in many cases have already caused) irreparable damage to the environment. They can cause even a global catastrophe, not only for reasons of technical inadequacy (technical failure of systems which are not well tested, not developed enough, hastily put in use, built without adequate security, etc.), but because of purely human factors like psychic stress, other psychic pathologies, and misperception for different reasons. Even accidents, like the recent disaster in Chernobyl, can give rise to these catastrophes. Very dangerous technologies, as well as scientific experiments, especially those done in vivo, should be legally, morally, and by all other means carefully controlled and, when necessary, forbidden. The faster the development of science and technology, the longer will need to be the list of the strictly prohibited activities.

While it is possible to control and to forbid ultra-hazardous technologies and other activities, the handling of the relevant knowledge itself is quite another question. It is evidently not easy nowadays to control much potentially dangerous knowledge, including the results of laboratory experiments, such as those done in vitro. Such knowledge, once attained, cannot of course be abolished, and is relatively easy, given place, means, and other circumstances (financial and otherwise) to apply in practice. This means that it could be relatively easily transformed into various means and techniques that could be used for military and other purposes hostile to the environment.

There are many questions in connection with this problem which are not easy to answer. Where, for example, does the freedom of scientific research end? Are all experiments In vitro acceptable? In short, where is the red line forbidding further tinkering with nature? It would be impossible, for instance, "to abolish nuclear knowledge without burning all books and all scientists. The prospect for that solution may have passed when the Pope failed to burn Galileo."

Duty of Disclosure

One of the possibilities is to lay greater stress on the duty of both states (i.e. through public authorities acting on their behalf) and scientific researchers themselves "to contribute to the definition of the aims and objectives of the programmes in which they are engaged and to the determination of the methods to be adopted which should be humanely, socially, and ecologically responsible." A very important issue in connection with this is the scientist's right to publish research results, including all results relevant to the state of the environment. People have the right to know the real state of the environment, and any attempt of state authorities, industrial enterprises, research institutions, or anybody else to keep vital information secret, or not to disclose it completely, is without doubt a violation of human rights. It is especially so when the information is of crucial importance for human existence.

Problems of Developing Countries

When it comes to the assessment and control of the development of science and technology, as well as environmental protection, developing countries are in a much worse position than the developed ones. First of all, as a rule, they do not have sufficient numbers of qualified scientists and other skilled personnel, and secondly they lack financial means and laboratory and other equipment. Besides, they do not co-operate sufficiently among themselves, in spite of all declarations to the contrary. In many developing countries the dangers arising from scientific and technological development are not per se regarded as a serious enough problem. Development is sometimes taken as a priority and an excuse for the massive impairment of the environment and violations of human rights. Many mistakes are made, including some of those already made by the developed countries long ago. Weak economies, unstable political systems, military adventures, nonexistent or bad environmental policies, unfavourable climate, the population "explosion," and many other factors only make the situation worse.

The developed countries, instead of increasing their efforts to help the developing countries in all possible ways to overcome their hardships and to make our world a better place for all, are often more interested in their own short-term interests. While giving some aid and financial assistance directly or through international organizations, they allow their firms, and especially some multinational corporations, to use the labour and natural resources of the third world without taking adequate measures for environmental protection. The work of the corporations, at least some of them, often impairs the environment even in the industrialized countries themselves. Their activities could be, as Capra wrote, "altogether disastrous in the third world. In those countries, where legal restrictions are often non-existent or impossible to enforce, the exploitation of people and of their land has reached extreme proportions."

It has been repeatedly stressed in many important international documents, conferences, and other fore, and by distinguished experts, that major changes in international economic and other closely interrelated domains are needed. The World Commission on Environment and Development believes, for instance, that: "Two conditions must be satisfied before international economic exchanges can become beneficial for all involved. The sustainability of the ecosystem on which the global economy depends must be guaranteed. And the economic partners must be satisfied that the basis of exchange is equitable; relationships that are unequal and based on dominance of one kind or another are not a sound and durable basis for interdependence. For many developing countries neither condition is met. " Both of the conditions should constitute basic elements of any new international economic order any other serious attempt to change international economic relations in general so as to better and improve the position of developing countries in particular. The aim of this article is not to review various proposals for the introduction of a new international economic order, or other ideas for improving the position of developing countries, but primarily to stress the importance of the ecological issue as a basic prerequisite for both the improvement of the economic situation in the world and for the better protection and promotion of human rights.

Disadvantaged Groups

Certain groups (ethnic minorities, old people, children, people with impaired health, women, the poorest strata of society, etc.) are often endangered more than the rest of the population and therefore deserve special attention. Their rights are violated in many ways, sometimes by subtle means that are hard to detect, sometimes openly and ruthlessly. Not having sufficient financial, political, and other power, and being in some ways already underprivileged and discriminated against, members of the minority groups are, for instance, forced to live in the most polluted parts of cities, are less educated, cannot buy good food and usually consume various kinds of cheap industrially prefabricated "junk food." They are often addiction-prone (narcotics, alcohol, tobacco), are not able to take various steps for countering the negative effects of the development of science and technology (by consuming health foods, by better health care, and in many other usually costly ways), and are not able to protect their rights by legal means, even when such means exist in their society. It is evident, however, that legal means are not sufficient seriously to improve their lot and that other ways and means must be sought, from better education to better health and other services. Probably more important than all these is improvement of their economic, social, and political position in society.

Technology Assessment

It is evident that one of the basic tasks in the future regulation of this probl-matique should be further development of appropriate machinery, on both international and national levels, for the assessment of the development of science and technology. That means an approach which is as democratic and, at the same time, as professional as possible. This is not easy to achieve. It is for governments to develop appropriate mechanisms for this assessment (taking into account their domestic, legal, political, and general situation). Governments, especially those of developed countries, are trying to do this by resorting increasingly to the engagement of science and technology advisers. Such advisers already play an important role, even in the foreign policy decision-making processes. At the same time traditional politicians and diplomats play an important role in the decision-making process on both the national and international scenes. The engagement of experts does not automatically solve all problems: in some cases the experts, as W.J.M. Mackenzie writes, "speak in fact for vested interests, or they may be totally at a loss and in hopeless disagreement." Politicians, for their part, are not, as a rule, expert in science and technology, and are influenced, among other things, by "the demands of political survival." All these factors, and many others, play a role in final decision-making in the field of science and technology.

Decisions concerning policy in the fields mentioned are at present usually made in the closed circles of various government offices, other official formal and informal groups and cliques, large industrial enterprises, etc. In most cases they are interested primarily in some kind of profit (financial, political, professional, etc.) while the interests of others, especially in the long term (other companies, other countries, population as a whole, mankind, future generations, etc.) are low on their priority lists. Technical lites, closely linked with military, industrial, political, and other circles of power, are often reluctant to disclose their activities. Besides that, the very nature of the development in most fields of science and technology prevents most people even from understanding what is going on. Who is to control scientists and technicians in their highly specialized fields? The old Roman maxim, *Quis custodies ipsos custodes*, becomes more than ever real. It is evident that further democratization of the decision- and policymaking processes is the best means of, on the one hard, protecting human rights from the negative influences which could flow from the development of science and technology, and, on the other, increasing the beneficial impact of this development.

The Right to Participate

The right to participate, including the right to be informed (to seek and receive information), in all activities concerning the development of science and technology, as well as economic development, environmental protection, and other cognate fields, is of the utmost importance. "The voluntary and democratic involvement of people in contributing to the development effort" by, among other things, taking part in decision-making, formulating policies, and controlling various activities, including those in the fields discussed in this article, is one of the ways to protect human rights from the negative influence of all kinds of deviations in policy-making. Various forms of participation, self-management systems, referendums, special parliamentary commissions, public opinion, the mass media (and especially a concerned and inquiring press), professional associations (with their codes of professional ethics), ecological, pacifists and other movements, consumer associations, as well as other formal and informal groups and individuals (including scientists in their role as conscientious people acting independently of the establishment), are only some of the possible ways of making the decision-making process as open and as democratic as possible, and of creating "a beneficial framework for individual and group self-realization, as well as for the healthy connection between state and society."

There are many problems in connection with the democratization of the decision-and policy-making process. It is true, especially in developed countries, that, generally speaking, "people are sufficiently well educated" to take part in the process of defining goals of scientific and technological development. In many cases, however, it is not easy to find experts impartial enough to judge and, if necessary, participate in the reassessment of a potentially dangerous technology or hazardous scientific experiment. This is especially the case in most developing countries, in which there are not enough experts even to properly assess imported foreign technology or chemicals, or to give advice on the extent to which an investment in industry would be environmentally sound. In fact, most developing, and in some fields even a number of developed countries, must rely on foreign expertise if they want to assess a technology or product, and since it is in the financial interest of foreign exporters - who come usually from developed but also from some developing countries - to show their otherwise polluting technology or unhealthy product in the best possible light, it happens that the experts asked to give their judgement are not always unbiased or entirely free of the possibility of bribery. One of the roles of international organizations, especially those belonging to the UN system, is to take part in the process of assessment, as well as to provide interested countries on request with all needed expertise for their development in science and technology and for environmental protection, in an impartial and professional way.

Humanitarian Law

The role of humanitarian law, as another autonomous branch of international law (much older, incidentally, than human rights law), should be given due regard. We share the opinion that "the links between human rights and humanitarian law are real and growing stronger." In certain circumstances these two complementary, although still distinct, systems become "convergent and perhaps interpenetrating." It is a fact that the International Committee of the Red Cross and the national societies of the Red Cross and Red Crescent, as well as the Henry Dunant Institute, are increasingly active in the field of human rights, not only in armed conflict but in areas such as health, education, the protection of victims of natural disasters, and the protection of the environment. It must be noted that serious disasters caused by human activities could be, under certain circumstances, regarded as a "public emergency which threatens the life of the nation," recognized in the International Covenant on Civil and Political Rights (Article 4). Cases like the Chernobyl accident, and other serious industrial accidents which have occurred in recent years, show that such a possibility is unfortunately not merely theoretical. Certain derogations of human rights "strictly required by the exigencies of the situation" could be expected, and international law should guarantee that the derogation does not last longer than necessary and that it does not infringe some of the basic (or sacrosanct) human rights. International co-operation should in such cases play an important role in both the protection of victims of the disaster and in giving proper assistance to the stricken country. Humanitarian law and international humanitarian organizations will certainly play an important role in such situations and the process of further convergence between human rights law and humanitarian law will continue.

Strengthening of Agencies

The strengthening and better organization of the existing UN bodies, and especially the Commission on Human Rights and the Human Rights Committee, is needed. The establishment of a permanent institution for assessing the impact of science and technology on human rights remains certainly one of the important steps towards better implementation of the steadily growing body of human rights law in general. It is necessary, however, that such a future institution, if established, devote special attention to development as "a comprehensive economic, social, cultural, and political process," as well as to environmental protection, which must naturally be done in co-operation with UNEP and other UN bodies.

Regional development deserves special attention, both from theoretical and practical points of view. Various groups of countries (West European, CMEA countries, American countries, etc.), linked by political, economic, territorial, and other ties, have been developing their own practices, as well as a theoretical base in the field of human rights, which certainly deserve further study and, whenever acceptable and desirable, transference to the universal level. Very interesting, indeed, have been developments on the basis of the European Convention for the Protection of Human Rights and Fundamental Freedoms (1950). In spite of the fact that the experiences of the West European countries, which share many political, economic, cultural, ideological, and other similarities, cannot be transferred automatically to others, they at least show that it is possible to improve international co-operation in the field of human rights by using legal means and procedures.

Finally, it is important to add that the experiences of various bilateral bodies, non-governmental organizations, national institutions and international, as well as national, professional associations could also make an important contribution to the development of the protection of human rights, as well as development in other fields, including the environmental issue, which are closely related to the human rights probl-matique.

ENVIRONMENTAL HARM AS A HUMAN RIGHTS

In a world of increasingly severe and wide-spread environmental degradation, new tools are needed to respond to environmental harm. Traditional international environmental law, that addresses the rights and obligations between nation-states, has little to offer individuals harmed by environmental damage. People whose health or livelihood is threatened by exposure to hazardous waste or the pollution of streams and rivers, for example, often have no recourse under international environmental laws. In addition, people harmed by environmental degradation are often ethnic minority groups, indigenous peoples, or people in occupied territories such as Tibet, who are marginalized within their own countries and effectively excluded from political participation or redress under national laws.

Linking human rights with the environment creates a rights-based approach to environmental protection that places the people harmed by environmental degradation at its center. Articulating the fundamental rights of peoples with respect to the environment creates the opportunity to secure those rights through human rights bodies in an international forum. This has particular force for the people most vulnerable to environmental harm and least able to access remedies or political support within their own countries. Connecting human rights and the environment reveals that human rights abuses often lead to environmental harm, just as environmental degradation may result in human rights violations.

The Environmental Dimension of Human Rights cannot be fully realized within a degraded or polluted environment. The fundamental right to life can be denied by events with environmental consequences: deaths caused by acute exposure to radioactivity or contaminated drinking water. The right to health; the right to safe and healthy working conditions; the right to adequate housing and food; these are all fundamental rights recognized in the Universal Declaration of Human Rights, and all have significant environmental dimensions. Exposure to toxic chemicals through careless hazardous waste disposal or industrial practices; the marginalization of pastoralists and subsistence farmers through soil depletion, deforestation, and confiscating traditional grazing lands; thesn plateau—fencing off traditional nomadic migration routes, or using traditional grazing lands for military installations or Chinese settlements—the result is a denial of fundamental human rights.

Denial of the human right to self-determination is also tightly linked to environmental degradation. Article 1 of the International Covenant on Civil and Political Rights—one of the fundamental international human rights instruments—declares unequivocally that "[a]ll peoples have the right of self-determination." Article 1 explains that "[b]y virtue of that right they freely determine their political status and freely pursue their economic, social and cultural development." The Covenant's right to self-determination includes environmental sovereignty, stated as a peoples' right to "freely dispose of their natural wealth and resources... [and] [i]n no case may a people be deprived of its own means of subsistence."

Denial of the right to self-determination is interwoven with the exploitation of a country's natural resources. The political and economic subjugation that exists in occupied territories creates social conditions under which extreme environmental degradation is unchecked. Deforestation of the Tibetan plateau through industrial clear-cutting by the Chinese has already claimed over one-half of the virgin forests of Tibet to feed China's enormous demand for wood and paper. It has also provided China with precious foreign exchange through exports of valuable old-growth logs to Japan. It is estimated that since the occupation of Tibet from 1959 to 1985, the benefit to China from the exploitation of Tibet's forests has been more than US$ 54 billion dollars.

Deforestation on the Tibetan plateau illustrates another potent aspect of connecting human rights and the environment. With respect to deforestation, it relates to the objective visibility of a hill-side of stumps. Clear-cuts are tangible—they can be photographed and counted—in a way that human rights violations taking place behind closed doors can not. This is particularly true in Tibet where foreign press and researchers are largely excluded, travel is restricted, and freedom of expression exists only on pain of imprisonment or worse. Describing environmental harm in human rights terms creates a measure of accountability for abuses that can otherwise be difficult to quantify, report, or even detect.

THE EFFECTS OF HUMAN RIGHTS VIOLATIONS ON THE ENVIRONMENT

Conditions in occupied Tibet demonstrate a prominent feature of environmental human rights. Environmental human rights encompass not only substantive protections such as the right to life, health, traditional livelihood, and culture, but they also include procedural rights. The procedural protections of environmental human rights include the right to information about the environment; the right to express opinions about environmental issues; the right to environmental and human rights education; the right to associate peacefully with others for the purpose of protecting the environment; the right to participate in decisions affecting the environment; and the right to administrative or judicial redress for violations of protected rights. These fundamental freedoms anchor the full realization of all human rights in an open and just society.

When Tibetans are systematically excluded from decisions affecting the Tibetan plateau's environment, and when Tibetans are not given information about actions that affect the environment nor permitted to exchange freely the information they have, there is no check on the environmental harm that results. The procedural protections of environmental human rights are as important as its substantive guarantees.

Uranium mining is taking place at nine known uranium mines on the Tibetan plateau to fuel China's nuclear weapons program and development of nuclear power. Due to the conditions of China's occupation, Tibetans are excluded from the decisions regarding where and how such mining is taking place, and have no access to information about the significant health risks associated with exposure to waste from the mining operations. As a result, rural Tibetan villages in the vicinity of uranium mines, which have experienced illness and "mysterious" deaths among villagers and domestic animals, still have no verifiable knowledge of what risks they are living with, and no recourse for the harms their communities have suffered.

The dual nature of environmental human rights not only calls attention to the environmental harm that has occurred, but also asks the question, "how did this happen?" By focusing on what decisions were made, who had access to information about them, and who took part in the decision, the social conditions under which environmental harm occurs can be better understood. Only when the people who are directly affected by environmental harm have a voice in what—and in what manner—environmentally significant actions are taken, is lasting environmental protection possible. This touches on one of the most innovative aspects of human rights and the environment. Linked in a rights-based approach, environmental human rights create a mechanism to hold polluters accountable for environmental harm. Giving the people personally affected by environmental harm a voice in an international forum creates accountability. It also melds one of the greatest strengths of human rights law to one of the weakest areas of environmental law: universal standards.

The large differences in levels of environmental protection between countries is one factor that is accelerating the exploitation of labour and natural resources around the world. As trans-national corporations and governments alike search for the cheapest means of production, the lack of environmental safeguards also means lower production costs. Low-priced goods sold on the international market often come at the expense of polluted air, fouled rivers, and deforested slopes. China has capitalized on Tibet's resource base to build part of its industrial infrastructure and place products on the world market. Similarly, one of China's most powerful foreign policy tools is its inexpensive labour pool and the role of its enormous population as a potential market for low-priced goods.

While the lack of universal standards is one of the biggest hurdles to environmental protection, this is one of the greatest strengths of human rights law. Fundamental notions of basic human rights have been articulated by the global community since 1948, when the Universal Declaration of Human Rights entered into force. Environmental human rights use global human rights norms to state a universal standard of minimum environmental protection that applies equally to every country. In this way environmental harm is cast in terms of its toll in human suffering, which leverages human rights standards to universalize our understanding of unacceptable environmental harm.

ENVIRONMENTAL HUMAN RIGHTS AS INTERNATIONAL LAW

These connections between human rights and the environment are evolving as international legal norms. In the twenty years between the United Nations (U.N.) Stockholm Declaration on the Human Environment in 1972 and the 1992 U.N. Conference on Environment and Development in Rio de Janeiro (the "Earth Summit"), increasing attention has been paid to the impact of environmental problems on human rights. Even in 1972, the Stockholm Declaration stated the human right to "adequate conditions of life, in an environment of a quality that permits a life of dignity and well-being." By 1992 with the Earth Summit's adoption of the Rio Declaration on Environment and Development, concepts of sustainable development and the rights of future generations to a healthy environment had been added: "environmental protection shall constitute an integral part of the development process and cannot be considered in isolation from it."

One of the most encouraging steps in this direction has been the work in the U.N. Sub-Commission on Prevention of Discrimination and Protection of Minorities over the last seven years. Since first taking up the issue in 1989, under the able leadership of the Sub-Commission's Special Rapporteur, the Sub-Commission has established that there exists meaningful precedent in international law for environmental human rights. The Sub-Commission explored the environmental dimension of human rights, and articulated the links between sustainable development, environmental protection and respect for human rights.

One of the most significant events of the Sub-Commission's study on human rights and the environment was a meeting of an international group of experts in May 1994 at the U.N. in Geneva, Switzerland. This meeting was called for by the Sub-Commission's Special Rapporteur and organized by the Sierra Club Legal Defence Fund, an American environmental non-governmental organization (NGO) that has played a role in the inception and progress of the Sub-Commission's work. This meeting had the ambitious goal of stating a set of principles that embodied the environmental dimension of human rights and grounded environmental rights firmly in human rights law.

The Geneva meeting achieved this goal, and produced the Draft Declaration of Principles on Human Rights and the Environment. The Draft Declaration is incorporated into the Sub-Commission's Final Report, and stands as the current reference point for environmental human rights.

ENVIRONMENTAL MOVEMENT

The environmental movement, a term that includes the conservation and green politics, is a diverse scientific, social, and political movement for addressing environmental issues. Environmentalists advocate the sustainable management of resources and stewardship of the environment through changes in public policy and individual behavior. In its recognition of humanity as a participant in (not enemy of) ecosystems, the movement is centered on ecology, health, and human rights.

The environmental movement is represented by a range of organizations, from the large to grassroots. Due to its large membership, varying and strong beliefs, and occasionally speculative nature, the environmental movement is not always united in its goals. At its broadest, the movement includes private citizens, professionals, religious devotees, politicians, and extremists.

History of the Movement

The roots of the modern environmental movement can be traced to attempts in nineteenth-century Europe and North America to expose the costs of environmental negligence, notably disease, as well as widespread air and water pollution, but only after the Second World War did a wider awareness begin to emerge.

The US environmental movement emerged in the late nineteenth and early twentieth century, with two key strands: protectionists such as John Muir wanted land and nature set aside for its own sake, while conservationists such as Gifford Pinchot wanted to manage natural resources for human use. Among the early protectionists that stood out as leaders in the movement were Henry David Thoreau, John Muir and George Perkins Marsh. Thoreau was concerned about the wildlife in Massachusetts; he wrote *Walden; or, Life in the Woods* as he studied the wildlife from a cabin. John Muir founded the Sierra Club, one of the largest conservation organizations in the United States. Marsh was influential with regards to the need for resource conservation.

Muir was instrumental in the creation of the world's first national park at Yellowstone in 1872.

During the 1950s, 1960s, and 1970s, several events illustrated the magnitude of environmental damage caused by humans. In 1954, the 23 man crew of the Japanese fishing vessel *Lucky Dragon 5* was exposed to radioactive fallout from a hydrogen bomb test at Bikini Atoll.

The publication of the book *Silent Spring* (1962) by Rachel Carson drew attention to the impact of chemicals on the natural environment. In 1967, the oil tanker *Torrey Canyon* went aground off the southwest coast of England, and in 1969 oil spilled from an offshore well in California's Santa Barbara Channel. In 1971, the conclusion of a law suit in Japan drew international attention to the effects of decades of mercury poisoning on the people of Minamata.

At the same time, emerging scientific research drew new attention to existing and hypothetical threats to the environment and humanity. Among them were Paul R. Ehrlich, whose book *The Population Bomb* (1968) revived concerns about the impact of exponential population growth. Biologist Barry Commoner generated a debate about growth, affluence and "flawed technology." Additionally, an association of scientists and political leaders known as the Club of Rome published their report *The Limits to Growth* in 1972, and drew attention to the growing pressure on natural resources from human activities.

Meanwhile, technological accomplishments such as nuclear proliferation and photos of the Earth from outer space provided both new insights and new reasons for concern over Earth's seemingly small and unique place in the universe. In 1972, the United Nations Conference on the Human Environment was held in Stockholm, and for the first time united the representatives of multiple governments in discussion relating to the state of the global environment. This conference led directly to the creation of government environmental agencies and the UN Environment Program. The United States also passed new legislation such as the Clean Water Act, the Clean Air Act, the Endangered Species Act, and the National Environmental Policy Act- the foundations for current environmental standards. By the mid-1970s anti-nuclear activism had moved beyond local protests and politics to gain a wider appeal and influence. Although it lacked a single co-ordinating organization the anti-nuclear movement's efforts gained a great deal of attention. In the aftermath of the Three Mile Island accident in 1979, many mass demonstrations took place. The largest one was held in New York City in September 1979 and involved two hundred thousand people; speeches were given by Jane Fonda and Ralph Nader. Since the 1970s, public awareness, environmental sciences, ecology, and technology have advanced to include modern focus points like ozone depletion, global climate change, acid rain, and the potentially harmful genetically modified organisms (GMOs).

SCOPE OF THE MOVEMENT

Environmental science is the study of the interactions among the physical, chemical and biological components of the environment.

- Ecology, or ecological science, is the scientific study of the distribution and abundance of living organisms and how these properties are affected by interactions between the organisms and their environment.

Primary Focus Points

The environmental movement is broad in scope and can include any topic related to the environment, conservation, and biology, as well as preservation of landscapes, flora, and fauna for a variety of purposes and uses. When an act of violence is committed against someone or some institution in the name of environmental defence it is referred to as eco terrorism

- The conservation movement seeks to protect natural areas for sustainable consumption, as well as traditional (hunting, fishing, trapping) and spiritual use.
- Environmental conservation is the process in which one is involved in conserving the natural aspects of the environment. Whether through reforestation, recycling, or pollution control, environmental conservation sustains the natural quality of life.
- Environmental health movement dates at least to Progressive Era, and focuses on urban standards like clean water, efficient sewage handling, and stable population growth. Environmental health could also deal with nutrition, preventive medicine, aging, and other concerns specific to human well-being. Environmental health is also seen as an indicator for the state of the environment, or an early warning system for what may happen to humans
- Environmental justice is a movement that began in the U.S. in the 1980s and seeks an end to environmental racism and prevent low-income and minority communities from an unbalanced exposure to highways, garbage dumps, and factories. The Environmental Justice movement seeks to link "social" and "ecological" environmental concerns, while at the same time preventing de facto racism, and classism. This makes it particularly adequate for the construction of labour-environmental alliances.

- Ecology movement could involve the Gaia Theory, as well as Value of Earth and other interactions between humans, science, and responsibility.
- Deep Ecology is an ideological spinoff of the ecology movement that views the diversity and integrity of the planetary ecosystem, in and for itself, as its primary value.
- Bright green environmentalism is a currently popular sub-movement, which emphasizes the idea that through technology, good design and more thoughtful use of energy and resources, people can live responsible, sustainable lives while enjoying prosperity.
- The anti-nuclear movement opposes the use of various nuclear technologies. The initial anti-nuclear objective was nuclear disarmament and later the focus began to shift to other issues, mainly opposition to the use of nuclear power. There have been many large anti-nuclear demonstrations and protests. Major anti-nuclear groups include Campaign for Nuclear Disarmament, Friends of the Earth, Greenpeace, International Physicians for the Prevention of Nuclear War, and the Nuclear Information and Resource Service.

ENVIRONMENTAL LAW AND THEORY

Property Rights

Many environmental lawsuits question the legal rights of property owners, and whether the general public has a right to intervene with detrimental practices occurring on someone else's land. Environmental law organizations exist all across the world, such as the Environmental Law and Policy Center in the midwestern United States.

Citizens' Rights

One of the earliest lawsuits to establish that citizens may sue for environmental and aesthetic harms was Scenic Hudson Preservation Conference v. Federal Power Commission, decided in 1965 by the Second Circuit Court of Appeals. The case helped halt the construction of a power plant on Storm King Mountain in New York State.

Nature's Rights

Christopher D. Stone's 1972 essay, "Should trees have standing?" addressed the question of whether natural objects themselves should have legal rights. In the essay, Stone suggests that his argument is valid because many current rightsholders (women, children) were once seen as objects.

ENVIRONMENTAL REACTIVISM

Numerous criticisms and ethical ambiguities have led to growing concerns about technology, including the use of potentially harmful pesticides, water additives like fluoride, and the extremely dangerous ethanol-processing plants.

NIMBY syndrome refers to public outcry caused by knee-jerk reaction to an unwillingness to be exposed to even necessary developments.

Some serious biologists and ecologists created the scientific ecology movement which would not confuse empirical data with visions of a desirable future world.

Modern Environmentalism

Today, the sciences of ecology and environmental science, rather than any aesthetic goals, provide the basis of unity to most serious environmentalists. As more information is gathered in scientific fields, more scientific issues like biodiversity, as opposed to mere aesthetics, are a concern. Conservation biology is a rapidly developing field. Environmentalism now has proponents in business: new ventures such as those to reuse and recycle consumer electronics and other technical equipment are gaining popularity. Computer liquidators are just one example.

In recent years, the environmental movement has increasingly focused on global warming as a top issue. As concerns about climate change moved more into the mainstream, from the connections drawn between global warming and Hurricane Katrina to Al Gore's film An Inconvenient Truth, many environmental groups refocused their efforts. In the United States, 2007 witnessed the largest grassroots environmental demonstration in years, Step It Up 2007, with rallies in over 1,400 communities and all 50 states for real global warming solutions.

Many religious organizations and individual churches now have programs and activities dedicated to environmental issues. The religious movement is often supported by interpretation of scriptures. Most major religious groups are represented including Jewish, Islamic, Anglican, Orthodox, Evangelical, Christian and Catholic.

Radical Environmentalism

Radical environmentalism emerged out of an ecocentrism-based frustration with the co-option of mainstream environmentalism. The radical environmental movement aspires to what scholar Christopher Manes calls "a new kind of environmental activism: iconoclastic, uncompromising, discontented with traditional conservation policy, at time illegal..." Radical environmentalism presupposes a need to reconsider Western ideas of religion and philosophy (including capitalism, patriarchy and globalization) sometimes through "resacralising" and reconnecting with nature.

Greenpeace represents an organisation with a radical approach, but has contributed in serious ways towards understanding of critical issues, and has a science-oriented core with radicalism as a means to mediaexposure. Groups like Earth First! take a much more radical posture.

Criticism

A study reported in *The Guardian* concluded that "people who believe they have the greenest lifestyles can be seen as some of the main culprits behind global warming."

The researchers found that individuals who were more environmentally conscious were more likely to take long-distance overseas flights, and that the resulting carbon emissions outweighed the savings from green lifestyles at home.

NOISE POLLUTION: NON-AUDITORY EFFECTS ON HEALTH

Noise, defined as 'unwanted sound', is perceived as an environmental stressor and nuisance. Non-auditory effects of noise, as dealt with in this chapter, can be defined as 'all those effects on health and well-being which are caused by exposure to noise, with the exclusion of effects on the hearing organ and the effects which are due to the masking of auditory information (*i.e.* communication problems).

Exposure to continuous noise of 85–90 dBA, particularly over a lifetime in industrial settings, can lead to a progressive loss of hearing, with an increase in the threshold of hearing sensitivity. Hearing impairments due to noise are a direct consequence of the effects of sound energy on the inner ear. However, the levels of environmental noise, as opposed to industrial noise, are much lower and effects on non-auditory health cannot be explained as a consequence of sound energy.

If noise does cause ill-health other than hearing impairment, what might be the mechanism? It is generally believed that noise disturbs activities and communication, causing annoyance.

In some cases, annoyance may lead to stress responses, then symptoms and possibly illness. Alternatively, noise may influence health directly and not through annoyance. The response to noise may depend on characteristics of the sound, including intensity, frequency, complexity of sound, duration and the meaning of the noise.

NON-AUDITORY EFFECTS OF NOISE ON HEALTH

Noise and Sleep Disturbance

There is both objective and subjective evidence for sleep disturbance by noise. Exposure to noise disturbs sleep proportional to the amount of noise experienced in terms of an increased rate of changes in sleep stages and in number of awakenings. Habituation occurs with an increased number of sound exposures by night and across nights. One laboratory study, however, found no habituation during 14 nights of exposure to noise at maximum noise level exposure. Objective sleep disturbance is likely to occur if there are more than 50 noise events per night with a maximum level of 50 dBA indoors or more. In fact, there is a low association between outdoor noise levels and sleep disturbance.

In the Civil Aviation Authority Study around Heathrow and Gatwick airports, the relative proportion of total sleep disturbance attributable to noise increased in noisy areas but not the level of total sleep disturbance. In effect, the work suggested a symptom reporting or attribution effect rather than real noise effects. In a subsequent actigraphy study around four UK airports, sleep disturbance was studied in relation to a wide range of aircraft noise exposure over 15 consecutive nights. Although there was a strong association between sleep EEGs and actigram-measured awakenings and self-reported sleep disturbance, none of the aircraft noise events were associated with awakenings detected by actigram and the chance of sleep disturbance with aircraft noise exposure of <82 dB was insignificant. Although it is likely that the population studied was one already adapted to aircraft noise exposure, this study is also likely to be closer to real life than laboratory studies with subjects newly exposed to noise. However, the actigraph as a sensitive measure of sleep disturbance has been questioned. Noise exposure during sleep may increase blood pressure, heart rate and finger pulse amplitude as well as body movements. There may also be after-effects during the day following disturbed sleep; perceived sleep quality, mood and performance in terms of reaction time all decreased following sleep disturbed by road traffic noise. Studies on noise abatement show that, by reducing indoor noise level, the amount of REM sleep and slow wave sleep can be increased. It thus seems that, although there may be some adaptation to sleep disturbance by noise, complete habituation does not occur, particularly for heart rate.

Noise Exposure and Performance

There is good evidence, largely from laboratory studies, that noise exposure impairs performance. Performance may be impaired if speech is played while a subject reads and remembers verbal material, although this effect is not found with non-speech noise. The effects of 'irrelevant speech' are independent of the intensity and meaning of the speech. The susceptibility of complex mental tasks to disruption by 'irrelevant speech' suggests that reading, with its reliance on memory, may also be impaired.

Perceived control over and predictability of noise has been found to be important in determining effects and after-effects of noise exposure. Glass and Singer found that tasks performed during noise were unimpaired but tasks that were carried out after noise had been switched off were impaired, this being reduced when subjects were given perceived control over the noise. Indeed, even anticipation of a loud noise exposure in the absence of real exposure may impair performance and an expectation of control counters this effect. Noise exposure may also slow rehearsal in memory, influence processes of selectivity in memory, and choice of strategies for carrying out tasks. There is also evidence that noise may reduce helping behaviour, increase aggression and reduce the processing of social cues seen as irrelevant to task performance.

NOISE AND CARDIOVASCULAR DISEASE

Noise exposure causes a number of predictable short-term physiological responses mediated through the autonomic nervous system. Exposure to noise causes physiological activation including increase in heart rate and blood pressure, peripheral vasoconstriction and thus increased peripheral vascular resistance. There is rapid habituation to brief noise exposure but habituation to prolonged noise is less certain.

The strongest evidence for the effect of noise on the cardiovascular system comes from studies of blood pressure in occupational settings. Many occupational studies have suggested that individuals chronically exposed to continuous noise at levels of at least 85 dB have higher blood pressure than those not exposed to noise. In many of these studies, noise exposure has also been an indicator of exposure to other factors, both physical and psychosocial, which are also associated with high blood pressure. Unless these other risk factors are controlled, spurious associations between noise and blood pressure may arise. A recent pioneering longitudinal industrial noise study has shown that noise levels predicted raised systolic and diastolic pressure in those doing complex but not simple jobs , and predicts increased mortality risk. Occupational noise exposure has also recently been linked to greater risk of death from motor vehicle injury. One possibility is that the effects of noise on blood pressure are mediated through an intermediate psychological response such as noise annoyance although this has not been convincingly proved.

Aircraft noise exposure around Schiphol Airport, Amsterdam has been related to more medical treatment for heart trouble and hypertension, more cardiovascular drug use and higher blood pressure, even after adjustment for age, sex, smoking, height/weight and socio-economic differences. The evidence of the effects of noise on coronary risk factors has not been especially consistent: effects of noise have been shown on systolic blood pressure (but not diastolic pressure), total cholesterol, total triglycerides , blood viscosity, platelet count and glucose level. However, a recent Swedish study found that the prevalence of hypertension was higher among people exposed to time-weighted energy averaged aircraft noise levels of at least 55 dBA or maximum levels above 72 dBA around Arlanda airport, Stockholm. In summary, there is some evidence from community studies that environmental noise is related to hypertension and there is also evidence that environmental noise may be a minor risk factor for coronary heart disease (Relative Risk 1.1–1.5).

A sudden intense exposure to noise may stimulate catecholamine secretion and precipitate cardiac dysrhythmias. However, neither studies in coronary care units of the effect of speech noise nor studies of noise from low altitude military flights on patients on continuous cardiac monitoring have detected changes in cardiac rhythm attributable to noise.

Exposure to high intensity noise in industry has been linked in some studies to raised levels of noradrenaline and adrenaline. In one study, catecholamine secretion decreased when workers wore hearing protection against noise. Some studies, but not all, have shown raised cortisol in relation to noise. The general pattern of endocrine responses to noise is indicative of noise as a stressor, exciting short-term physiological responses, but there are inconsistencies between studies.

It has been postulated that noise exposure creates annoyance which then leads on to more serious psychological effects. This pathway remains unconfirmed; rather it seems that noise causes annoyance and, independently, mental ill-health also increases annoyance. A more complex model incorporates the interaction between the person and their environment. In this model, the person readjusts their behaviour in noisy conditions to reduce exposure. An important addition is the inclusion of the appraisal of noise (in terms of danger, loss of environmental quality, meaning of the noise, challenges for environmental control, *etc.*) and coping (the ability to alter behaviour to deal with the stressor). This model emphasizes that dealing with noise is not a passive process.

Symptoms reported among industrial workers regularly exposed to high noise levels in settings such as schools and factories include nausea, headaches, argumentativeness and changes in mood and anxiety. Many of these industrial studies are difficult to interpret, however, because workers were exposed to other stressors such as physical danger and heavy work demands, in addition to excessive noise. Community surveys have found that high percentages of people reported 'headaches', 'restless nights', and 'being tense and edgy' in high-noise areas ,. An explicit link between aircraft noise and symptoms emerging in such studies raised the possibility of a bias towards over-reporting of symptoms. Notably, a study around three Swiss airports , which did not mention that it was related to aircraft noise, did not find any association between the level of exposure to aircraft noise and symptoms.

Noise and Common Mental Disorder

Early studies found associations between the level of aircraft noise and psychiatric hospital admission rates both in London and Los Angeles , but this has not been convincingly confirmed by more recent studies. In community studies such as the West London Survey of Psychiatric Morbidity , no overall relationship was found between aircraft noise and the prevalence of psychiatric morbidity using various indices of noise exposure. In longitudinal analyses in the Caerphilly Study, no association was found between road traffic noise and psychiatric disorder, even after adjustment for socio-demographic factors and baseline psychiatric disorder, although there was a small non-linear association of noise with increased anxiety scores.

Some studies have found dose–response associations: exposure to higher levels of military aircraft noise around Kadena airport in Japan was related in a dose–response relationship to depressiveness and nervousness , and road traffic noise has been weakly associated with mental health symptoms after adjusting for age, sex, income and length of residence. Overall, environmental noise seems to be linked to psychological symptoms but not to clinical psychiatric disorder. However, there may be a link to psychiatric disorder at much higher noise levels.

The most widespread and well documented subjective response to noise is annoyance, which may include fear and mild anger, related to a belief that one is being avoidably harmed. Noise is also seen as intrusive into personal privacy, while its meaning for any individual is important in determining whether that person will be annoyed by it.

Annoyance reactions are often associated with the degree of interference that any noise causes in everyday activities, which probably precedes and leads on to annoyance. In both traffic and aircraft noise studies, noise levels have been found to be associated with annoyance in a dose–response relationship ,. Overall, it seems that conversation, watching television or listening to the radio (all involving speech communication) are the activities most disturbed by aircraft noise while traffic noise, if present at night, is most disturbing for sleep.

One of the primary characteristics affecting the unwantedness of noise is its loudness or perceived intensity. Loudness comprises the intensity of sound, the tonal distribution of sound and its duration. The evidence is mixed on the importance of both the duration and the frequency components of sound and also the number of events involved in determining annoyance. High frequency noise has been found to be more annoying than low frequency noise. Vibrations are perceived as a complement to loud noise in most community surveys of noise and are found to be important factors in determining annoyance, particularly because they are commonly experienced through other senses as well as hearing. Fields found that, after controlling for noise level, noise annoyance increases with fear of danger from the noise source, sensitivity to noise, the belief that the authorities can control the noise, awareness of the non-noise impacts of the source and the belief that the noise source is not important.

Combined Effects of Noise Exposure and other Stressors

Noise effects on health may be augmented by, or in turn may augment, the impact of other stressors on health. Stressors may act synergistically, antagonistically or not at all. Stressors may include physical, chemical, biological, social and work organizational factors. In a laboratory based experiment, an interaction was found between having a cold and noise exposure on simple reaction time. There was little difference between healthy and cold subjects' performance tested in quiet conditions, but for subjects tested in noisy conditions (70 dBA), performance was much slower for the cold subjects. Synergistic effects of exposure to noise and vibration have been demonstrated on diastolic blood pressure, whereas temperature and noise have been shown to affect morning adrenaline secretion.

There has been much emphasis on laboratory studies without considering that results of such studies may lack external validity. Past research on combined effects has not considered common conditions and levels of stressors across studies, direct and indirect effects, long durations of exposure and complex tasks. Field studies suggest that the effects of multiple stressors have greater combined effects than simply summing individual stressors. Few field studies have examined the effects of multiple environmental stressors. This could be an important new area for the development of noise research.

It is likely that children represent a group which is particularly vulnerable to the non-auditory health effects of noise. They have less cognitive capacity to understand and anticipate stressors and lack well-developed coping strategies. Moreover, in view of the fact that children are still developing both physically and cognitively, there is a possible risk that exposure to an environmental stressor such as noise may have irreversible negative consequences for this group.

Studies of children exposed to environmental noise have consistently found effects on cognitive performance. The studies which are most informative in terms of the effects of noise on cognition have been field studies focusing on primary school children. This article will focus on these studies.

The effects of noise have not been found uniformly across all cognitive functions. The research evidence suggests that chronic exposure to noise affects cognitive functions involving central processing and language comprehension. The effects which have been found can be summarized as follows. Deficits have been found in sustained attention and visual attention. Relatedly, according to teachers' reports, noise-exposed children have difficulties in concentrating in comparison with children from quieter schools. Children exposed to chronic environmental noise have been found to have poorer auditory discrimination and speech perception as well as poorer memory requiring high processing demands. Finally, chronically exposed children tend to have poorer reading ability and school performance on national standardized tests.

The first well-designed naturalistic field study to examine the effects of chronic noise exposure focused on primary school children living in four 32-floor apartment buildings adjacent to a major road.

The rationale behind this study was that children in the lower floor of the apartment building would be exposed to higher amounts of noise from the road than those higher up the building. Seventy-three children were tested for auditory discrimination and reading level and the results indicated that children living on the lower floors had greater impairments on these measures than those living higher up the buildings.

A very well controlled study by Bronzaft and McCarthy compared primary school children taught in a classroom which was exposed to high levels of railway noise with children in a quiet classroom in the same school. Significant differences in reading scores were found between children in the two classrooms. In fact, the mean reading age of the noise-exposed children was 3–4 months behind that of the control children.

A series of studies have been carried out in schools around Heathrow Airport in west London. These studies have used repeated-measures designs to compare noise-exposed and control children. In the first of these studies , the cognitive performance and stress responses of 9- to 10-year-old children in four high noise schools were compared with those of children in four matched control schools. The results showed that, at baseline, the noise-exposed children had impaired reading comprehension and sustained attention after adjustment for age, main language spoken at home and social deprivation. The results at follow up 1 year later suggest that the children's further development in reading comprehension may be affected.

The second study to be conducted near Heathrow Airport was a multi-level modelling study of national standardized test scores (SATs). The data for 11,000 eleven-year-old children were analysed in relation to aircraft noise exposure contours. The results showed that noise exposure was associated with performance on reading and maths tests in a dose–response function but that this was influenced by socio-economic factors.

The most recent study to be carried out at Heathrow compared the cognitive performance and stress responses of children in 10 high-noise schools with those of children in 10 matched control schools. The results indicated that children in the noise-exposed schools experienced greater annoyance and had poorer reading performance on the difficult items of a national standardized reading test.

Perhaps the most important of all the naturalistic field studies to examine the effects of noise exposure on children was that carried out in Munich in the 1990s.

This prospective, longitudinal study was able to take advantage of a naturally occurring experiment in which the existing Munich Airport was closed down and a new airport was opened at another location. Data were collected at both sites across three testing waves, one before the closure of the old airport and opening of the new one and two afterwards.

The mean age of children was 10.8 years. The cross-sectional results showed that, at Wave 1, children at the old airport displayed effects on long-term episodic memory and reading comprehension. The longitudinal results showed that after three waves of testing, children at the old airport had improvements in long-term memory, suggesting that this effect of noise exposure is reversible.

Interestingly, by the third wave of testing children at the new airport were exhibiting deficits in long-term memory and reading comprehension, providing strong evidence for a causal link between noise exposure and cognitive effects.

A number of studies have identified an association between chronic exposure to aircraft noise and decreased motivation. The results are however not consistent. In the Los Angeles Airport Study children exposed to chronic aircraft noise were less likely to solve a difficult puzzle involving a success or failure experience and were more likely to give up. In a follow-up 1 year later[78] the finding that noise-exposed children were less likely to solve a difficult puzzle was replicated, but the finding that the same children are more likely to give up on a difficult puzzle was not. In the Munich study , noise-exposed children gave up on an insoluble puzzle more quickly than their non noise-exposed counterparts.

Cardiovascular Effects

In addition to effects on cognitive performance, there is evidence that chronic noise exposure may give rise to physiological effects in terms of raised blood pressure. In the Los Angeles Airport Study , chronic exposure to aircraft noise was found to be associated with raised systolic and diastolic blood pressure.

These increases, although significant, were within the normal range and were not indicative of hypertension. At follow-up 1 year later , the findings were the same, showing that these effects had not habituated.

In the Munich study, chronic noise exposure was found to be associated with both baseline systolic blood pressure and lower reactivity of systolic blood pressure to a cognitive task presented under acute noise. After the new airport opened, a significant increase in systolic blood pressure was observed providing evidence for a causal link between chronic noise exposure and raised blood pressure.

No association was found between noise and diastolic blood pressure or reactivity.

The Munich Airport Study examined overnight, resting levels of urinary catecholamines (adrenaline and noradrenaline). In the cross-sectional study at the old airport, endocrine levels were significantly higher in the noise-exposed children, indicating raised stress levels. The longitudinal data reveal a sharp increase in catecholamine levels in noise-exposed children following the opening of the new airport. Cortisol levels were also examined but no significant differences were observed in either the cross-sectional or longitudinal data. This latter finding is consistent with that of one of the Heathrow Studies.

Studies have consistently found evidence that exposure to chronic environmental noise causes annoyance in children, even in young children. In Munich, noise-exposed children were found to be more annoyed by noise as indexed by a calibrated community measure. In London, child-adapted, standard self-report questions were used to assess annoyance and showed higher annoyance levels in noise-exposed children. In a follow-up 1 year later, the same result was found, suggesting that annoyance effects are not subject to habituation.

The evidence for effects of environmental noise on health is strongest for annoyance, sleep and cognitive performance in adults and children. Occupational noise exposure also shows some association with raised blood pressure. Dose–response relationships can be demonstrated for annoyance and, less consistently, for blood pressure.

The effects of noise are strongest for those outcomes that, like annoyance, can be classified under 'quality of life' rather than illness. What these effects lack in severity is made up for in numbers of people affected, as these responses are very widespread.

It may be that the risk of developing mental or physical illness attributable to environmental noise is quite small, although it is too soon to be certain of this in terms of the progress of research. Part of the problem is that the interaction between people, noise and ill-health is a complex one. Humans are not usually passive recipients of noise exposure and can develop coping strategies to reduce the impact of noise exposure. If people do not like noise they may take action to avoid it by moving away from noisy environments or, if they are unable to move away, by developing coping strategies. Active coping with noise may be sufficient to mitigate any ill-effects. Perception of control over the noise source may reduce the threat of noise and the belief that it can be harmful. It may also be that noise is more harmful to health in situations where several stressors interact and the overall burden may lead to chronic sympathetic arousal or states of helplessness.

Adaptation to long-term noise exposure needs further study. Most people exposed to chronic noise, for instance from major airports, seem to tolerate it. Yet, questionnaire studies suggest that high levels of annoyance do not decline over time. Another possibility is that adaptation to noise is only achieved with a cost to health. Evans and Johnson found that maintaining task performance in noisy offices was associated with additional physiological effort and hormonal response.

Undoubtedly, there is a need for further research to clarify this complex area, including better measurement of noise exposure and health outcomes. Moreover, there should be a greater emphasis on field studies using longitudinal designs with careful choice of samples to avoid undue bias related to prior noise exposure.

5

The Struggle over Environmental Human Rights

In October 2000, murmurs of horror began to do the rounds of the world's press when the Indian writer Arundhati Roy, author of *The God of Small Things*, had to serve a day in prison for contempt of court. A few weeks earlier, with righteous anger she had appeared before the media and publicly scolded India's Supreme Court. Her outrage was directed at a judgement that had spelt disaster for the country's largest ecological movement, India's holiest river and tens of thousands of small farmers: green light for the giant Narmada River dam project to go ahead! In the eyes of Roy and large sections of the Indian public, this raised to new heights the state's arrogant contempt for ordinary people as well as for economic rationality.

She wrote: In India over the last ten years the fight against the Sardar Sarovar Dam has come to represent far more than the fight for one river. ... From being a fight over the fate of a river valley it began to raise doubts about an entire political system. What is at issue now is the very nature of our democracy. Who owns this land? Who owns its rivers? Its forests? Its fish? These are huge questions. They are being taken hugely seriously by the State. They are being answered in one voice by every institution at its command – the army, the police, the bureaucracy, the courts. And not just answered, but answered unambiguously, in bitter, brutal ways (Roy, 1999).

Who has the advantages and who the disadvantages in the use of nature? This is the key question of environmental justice. It starts from the observation that often benefits and burdens of resource use do not accrue to one and the same social player, but are unevenly spread across different social groups. What economists like to call 'the externalization of negative consequences' has not only a biophysical but also a social profile: that is to say, the mechanism whereby benefits are internalized and costs externalized structures societies into winners and losers. It is a mechanism driven by an asymmetry of power. Power relations establish a social differential, which ensures that positive consequences crystallize at the top end and negative consequences at the bottom end. This shifting of costs may take place in a temporal, spatial or social dimension: that is, costs may be shifted temporally from present to future, spatially from centre to periphery, and socially from upper classes to lower classes.

The mass demonstrations in the Narmada valley, with Arundhati Roy in the middle, can be considered as a struggle over environmental human rights. For at a certain degree of vulnerability, burdens tip over into injustice and threaten the basic rights of the people involved. When injustice of this kind arises out of ecological disadvantages, it throws up a number of human rights issues. For resource flows do not come only with an ecological baggage in the form of an indirect consumption of materials. In some circumstances, they also come with a social baggage in the form of human rights violations.

Resources and Subsistence Rights

If people do not have the basic capabilities to support themselves in dignity, their human rights are under threat. In particular, nourishment, health, housing and livelihood may be thought of as subsistence rights (Shue, 1980): that is, as requirements for a minimum degree of economic security. Subsistence rights, then, are part of human rights; they form the core rights of the International Covenant on Economic, Social and Cultural Rights.

Environmental human rights come into play when subsistence is dependent on the right to use natural spaces. For in addition to income and community ties, nature is another important source of livelihood, especially for the third of humanity who rely on direct access to its fields, forests, pastures and waters (UNDP, 1998: 80). For people who live in a process of unmediated exchange with nature, the fate of ecosystems in the area they inhabit is a question of vital importance. Their subsistence rights depend to a considerable extent upon the availability of natural resources. Therefore, any degradation of the natural spaces on which they depend for a living undermines their subsistence rights.

Again and again, however, the natural living space of the poor comes into the firing line of the international resource economy. The peripheries dotted around the globe, in both rural and urban areas, are not beyond the reach of the world economy, but have many different ties to the core states precisely in respect of natural resources. They may serve as: (1) a hinterland for the extraction of raw materials, (2) a source of agricultural produce, or

(3) a deployment area for genetically modified organisms. And they may prove to be (4) danger zones because of climate change, (5) foci of disease because of pollution or (6) an arena of marginalization because of the evolution of resource prices.

Generally, geographical distance and/or huge societal differences separate the peripheries from the core zones that place their resource situation under strain; the locations of enrichment are mostly at a safe remove from the locations of impoverishment. And yet investment-flows, atmospheric chemistry or the pull of higher purchasing power create bonds within a power gradient that ensures a stable asymmetry of advantages and disadvantages. Resource conflicts therefore break out in the poor countries of the periphery, where the struggle centres, if only at the local level, on nothing less than rights to the biosphere. By virtue of man's biological nature, some of these rights are inalienable. If they cannot be guaranteed, a conflict over resources turns into a conflict over human rights.

Conflicts over the Extraction of Raw Material

Ever since the age of Pizarro, the 'New World' has been combed for valuable raw materials. But today the exploration and exploitation of new sources stretches into the remotest parts of the world's sea and land masses. Oil is extracted from deep inside the tropical forest and deep beneath the ocean waves; timber is carried from faraway Patagonia and Siberia; and floating fish factories plough the seas from the Arctic to the Antarctic. The opening of frontiers to foreign corporations has intensified the pressure to move forward the front line of exploitation.

However, oil and mineral extraction often damages not only plant and animal biotopes but human communities as well, especially those of indigenous peoples. In fact, it is the lands of the original inhabitants that are now being caught up in the worldwide flow of resources. Their landscapes are often degraded, ravaged and subject to sacrilege. Gold mines in Romania, Ghana or South Africa, as well as silver and tin mines in Peru or Chile, have extensively poisoned the rivers. And the development of oil fields has come at a price for the forest biotopes of Colombia, Brazil, Peru and Ecuador, in the Niger delta as well as the Siberian tundra.

Conflicts over the Alteration of Ecosystems

The extraction of raw materials is only one of the ways in which humans help themselves to resources from the biosphere; the conversion of ecosystems into production machines is the other. For this, the land, climate and range of species in the southern hemisphere often provide a combination of local conditions that are scarcely to be found anywhere in the North. The history of colonization, up to and including the age of globalization, is therefore also a history of the occupation and settlement of land. From tea and sugar cane through cotton and eucalyptus trees to kiwi fruit and prawns, agrarian systems have been constructed to fill the tables of distant consumers. The resource conflict between subsistence economies and market economies is one of the roots of the present clashes over the modification of nature for plantations, aquaculture and water reservoirs.

Conflicts over Genetic Engineering

Along with extraction and colonization, a third way of using nature as a resource has recently made its appearance: namely, the reprogramming of nature itself. The spread of biotech corporations into the agricultural regions of the South poses a new threat to subsistence rights. From now on, certain kinds of seeds and economically useful plants and animals may come with a hefty price-tag attached, since patent-protected property rights cover genetically modified life-forms. Once, however, a growing number of economically useful life-forms come under the ownership of corporations, access to the means of gaining a livelihood becomes restricted: farmers who previously exchanged seeds, collected shoots or reproduced livestock free of charge now have to pay license fees to use the patrimony of nature.

Conflicts over Resource Prices

The fate of a household with too little money may be related to natural resources in one of two ways: either through production prices or through consumption prices. In the first case, very small-scale farmers are only just able to scrape a living by selling their produce on the local or national market. Their vulnerability is based on the instability of market prices, so that an unexpected collapse of the price for their beans, oil crops, nuts, etc. (perhaps due to cheaper imports from abroad) may plunge them into an existential crisis. In the second case, price rises cut off poor sections of the population (mostly living in shanty towns) from a number of essential environmental goods. Recent conflicts in this field have mainly centred on water or electricity, since the privatization of public services in the big cities of the South means that these goods too have become commodities to be sold at a profit. This threatens the subsistence rights of the poor. As events in Cochabamba, Soweto and Jakarta have shown, a huge potential for conflict builds up when unjust access to resources deprives sections of the population of the source of all life: water.

Subsistence Rights and Human Rights

Before the Second World War, it was only states that could claim rights. The rights of persons were first recognized at the international level only with the Universal Declaration of Human Rights in 1948. This may be seen as the juridical revolution in human rights (Ignatieff, 2001), which went together with a revolution in their advocacy and enforcement. For a long time, however, people's economic, social and cultural rights played a subordinate role in this growing legal awareness. This was also due to the Cold War: for the Western bloc inscribed civil and political rights on its banner, while the Eastern bloc did the same with economic and social rights. The two sets of human rights were ritually played off against each other, with the result that social rights were taken no more seriously in the West than democratic rights were in the East. Meanwhile, this confrontation has resolved itself, and the inseparability and interdependence of human rights has been largely accepted (Steiner and Alston, 1996). Indeed, it would be hard to understand why disease or inadequate nourishment should be less important than press censorship or religious persecution in affecting people's ability to act. Without social and economic rights, the minimum basis is lacking for equality of civil and political rights – and, conversely, social and economic rights without civil and political rights are robbed of the motive power of freedom. A minimalist conception of human rights that refers only to negative freedoms therefore discriminates against the have-nots and those whose livelihood is threatened; recognition of their dignity requires the protection of subsistence rights.

Now and then it has been claimed that civil and political rights only create a duty to refrain from certain actions, whereas economic and social rights require positive action with the help of appropriate resources. This distinction between negative and positive rights may be overcome (Steiner and Alston, 1996: 282). For negative rights also sometimes require positive action if their validity is to be ensured: for example, freedom from censorship is meaningful only if press pluralism is guaranteed. Conversely, a policy of refraining from certain actions is important in relation to positive rights: for example, the right to adequate food implies that action will not be taken that leads to the expropriation of land on which the food is grown. In the case of all rights, therefore, it is meaningful to distinguish three levels of obligation on a scale from negative to positive: the obligation to respect, to protect and to guarantee.

A human rights policy, therefore, involves not only negative but also positive rights, not only checks on the state but also deployment of the state; and that human rights may be not only violated but also withheld. On the other hand, a maximalist conception of human rights is also misleading. Everyone has a right to drinking water, but not to a heart operation. The desirable must be distinguished from the necessary, and the successful from the legally recoverable. 'Human rights is an account of what is right, not of what is good' (Ignatieff, 2001: 55). Political goals do not constitute rights; one may have no obligation to the former but a definite obligation to the latter. But the more that political goals are claimed to be obligations, the closer becomes a collision between everyone's livelihood right and everyone's right to freedom and cultural diversity. For this reason it is advisable, as in the liberal tradition, to formulate respect for subsistence rights first of all in terms of negative right, social institutions should be shaped in such a way that they do not structurally and permanently undermine fundamental rights (Pogge, 2002).

Climate Change and Livelihood Rights

This time, colonial destruction comes without imperial power and without an army of occupation. Instead it advances through the air, invisible and insidious, transported over long distances by the chemistry of the atmosphere. When the earth's atmosphere grows warmer, nature becomes unstable. Many biotopes thus become less inhabitable or, in extreme cases, actually uninhabitable for certain kinds of plants, animals and even humans. Obviously, a rise in sea level will make some of the most densely populated areas of the globe impossible to live in. Less evident is the fact that changes in humidity and temperature will trigger changes in vegetation, species diversity, soil fertility and water deposits – not to speak of possible natural disasters. But the risks from global warming are by no means evenly distributed across the world's population; they disproportionately affect the socially weak and powerless, who already scrape a bare existence on marginal lands and in impoverished city outskirts.

Environmental Degradation in Cities

Whereas the rural poor are often denied access to the natural resources needed for their survival, urban slum-dwellers are more physically threatened by the decay of their immediate environment. They cannot rely upon the services of nature that are essential to them as biological beings. Non-monetary goods such as clean air, water, shelter and security are less available in cities than in the countryside. Along with their lack of money, the urban poor have to contend with polluted water, unsafe housing, dirty air and a high crime-rate. They almost never have access to safe, healthy and spacious accommodation, to legally watertight rental contracts or property deeds or to dependable services and facilities; they frequently live in parts of town where the first cloudburst may trigger a mudslide. It is therefore not surprising that a close correlation can be established between income level and environmental risk.

POLLUTION EFFECTS ON HUMANS, ANIMALS, PLANTS AND THE ENVIRONMENT

There is no doubt that excessive levels of pollution are causing a lot of damage to human & animal health, plants & trees including tropical rainforests, as well as the wider environment. All types of pollution – air, water and soil pollution – have an impact on the living environment.

The effects in living organisms may range from mild discomfort to serious diseases such as cancer to physical deformities; ex., extra or missing limbs in frogs. Experts admit that pollution effects are quite often underestimated and that more research is needed to understand the connections between pollution and its effects on all life forms.

ENVIRONMENTAL POLLUTION EFFECTS ON HUMANS

We know that pollution causes not only physical disabilities but also psychological and behavioral disorders in people. We are discussing the effects of air pollution and specific air pollutants in more detail in the Air Pollutants article. The following pollution effects on humans have been reported:

Air Pollution Effects

- Reduced lung functioning
- Irritation of eyes, nose, mouth and throat
- Asthma attacks
- Respiratory symptoms such as coughing and wheezing.
- Increased respiratory disease such as bronchitis
- Reduced energy levels
- Headaches and dizziness
- Disruption of endocrine, reproductive and immune systems
- Neurobehavioral disorders
- Cardiovascular problems
- Cancer
- Premature death.

Water Pollution Effects

Waterborne diseases caused by polluted drinking water:

- Typhoid
- Amoebiasis
- Giardiasis
- Ascariasis
- Hookworm.

Waterborne diseases caused by polluted beach water:

- Rashes, ear ache, pink eye
- Respiratory infections.
- Hepatitis, encephalitis, gastroenteritis, diarrhoea, vomiting, and stomach aches.

Conditions related to water polluted by chemicals (such as pesticides, hydrocarbons, persistent organic pollutants, heavy metals etc):

- Cancer, incl. prostate cancer and non-Hodgkin's lymphoma
- Hormonal problems that can disrupt reproductive and developmental processes
- Damage to the nervous system
- Liver and kidney damage
- Damage to the DNA
- Exposure to mercury (heavy metal):
    - o *In the womb:* may cause neurological problems including slower reflexes, learning deficits, delayed or incomplete mental development, autism and brain damage
    - o *In adults:* Parkinson's disease, multiple sclerosis, Alzheimer's disease, heart disease, and even death.

Other notes:

- Water pollution may also result from interactions between water and contaminated soil, as well as from deposition of air contaminants (such as acid rain)
- Damage to people may be caused by fish foods coming from polluted water (a well known example is high mercury levels in fish)
- Damage to people may be caused by vegetable crops grown / washed with polluted water (author's own conclusion).

Soil Pollution Effects

- Causes cancers including leukaemia
- Lead in soil is especially hazardous for young children causing developmental damage to the brain

- Mercury can increase the risk of kidney damage; cyclodienes can lead to liver toxicity
- Causes neuromuscular blockage as well as depression of the central nervous system
- Also causes headaches, nausea, fatigue, eye irritation and skin rash.

Other notes:

- Contact with contaminated soil may be *direct* (from using parks, schools etc) or *indirect* (by inhaling soil contaminants which have vaporized)
- Soil pollution may also result from secondary contamination of water supplies and from deposition of air contaminants (for example, via acid rain)
- Contamination of crops grown in polluted soil brings up problems with food security
- Since it is closely linked to water pollution, many effects of soil contamination appear to be similar to the ones caused by water contamination.

An Extreme Oil Pollution Case

Pollution of pristine Ecuador rainforest by Texaco / Chevron oil corporation represents perhaps one of the most outrageous cases of oil pollution ever.

Some levels of pollutants left by the company on its sites of oil exploration have been calculated to exceed the US safety standards by as much as 1,000 times, causing such side effects as children born with fused fingers and deformed eyes, high cancer rates, etc.

ENVIRONMENTAL POLLUTION EFFECTS ON ANIMALS

- *Acid rain* (formed in the air) destroys fish life in lakes and streams
- Excessive *ultraviolet radiation* coming from the sun through the ozone layer in the upper atmosphere which is eroded by some air pollutants, may cause skin cancer in wildlife
- *Ozone* in the lower atmosphere may damage lung tissues of animals.

Effects of Pollution on Animals - Water Pollution

- *Nutrient pollution* (nitrogen, phosphates etc) causes overgrowth of toxic algae eaten by other aquatic animals, and may cause death; nutrient pollution can also cause outbreaks of fish diseases
- *Chemical contamination* can cause declines in frog biodiversity and tadpole mass
- *Oil pollution* (as part of chemical contamination) can negatively affect development of marine organisms, increase susceptibility to disease

and affect reproductive processes; can also cause gastrointestinal irritation, liver and kidney damage, and damage to the nervous system

- *Mercury* in water can cause abnormal behavior, slower growth and development, reduced reproduction, and death
- *Persistent organic pollutants* (POPs) may cause declines, deformities and death of fish life
- Too much sodium chloride (ordinary salt) in water may kill animals.

Other notes:

- We also assume that some higher forms of non-aquatic animals may have similar effects from water pollution as those experienced by humans, as described above.

Effects of Pollution on Animals - Soil Pollution

- Can alter metabolism of microorganisms and arthropods in a given soil environment; this may destroy some layers of the primary food chain, and thus have a negative effect on predator animal species
- Small life forms may consume harmful chemicals which may then be passed up the food chain to larger animals; this may lead to increased mortality rates and even animal extinction.

ENVIRONMENTAL POLLUTION EFFECTS ON TREES AND PLANTS

Air Pollution

- *Acid rain* can kill trees, destroy the leaves of plants, can infiltrate soil by making it unsuitable for purposes of nutrition and habitation
- *Ozone holes* in the upper atmosphere can allow excessive ultraviolet radiation from the sun to enter the Earth causing damage to trees and plants
- *Ozone* in the lower atmosphere can prevent plant respiration by blocking stomata (openings in leaves) and negatively affecting plants' photosynthesis rates which will stunt plant growth; ozone can also decay plant cells directly by entering stomata.

Water Pollution

- May disrupt photosynthesis in aquatic plants and thus affecting ecosystems that depend on these plants
- Terrestrial and aquatic plants may absorb pollutants from water (as their main nutrient source) and pass them up the food chain to consumer animals and humans

- Plants may be killed by too much sodium chloride (ordinary slat) in water
- Plants may be killed by mud from construction sites as well as bits of wood and leaves, clay and other similar materials
- Plants may be killed by herbicides in water; herbicides are chemicals which are most harmful to plants.

Soil Pollution

- May alter plant metabolism and reduce crop yields
- Trees and plants may absorb soil contaminants and pass them up the food chain.

ENVIRONMENTAL POLLUTION EFFECTS ON WIDER ENVIRONMENT

Apart from destroying the aquatic life in lakes and streams, acid rain can also corrode metals, damage surfaces of buildings and monuments, and cause soil acidification. Pollution of water may cause oxygen depletion in marine environments and severely affect the health of whole ecosystems.

ENVIRONMENTAL POLLUTION

Its Sources and Effects

Environmental pollution is "the contamination of the physical and biological components of the earth/atmosphere system to such an extent that normal environmental processes are adversely affected". "Pollution is the introduction of contaminants into the environment that cause harm or discomfort to humans or other living organisms, or that damage the environment" which can come "in the form of chemical substances, or energy such as noise, heat or light". "Pollutants can be naturally occurring substances or energies, but are considered contaminants when in excess of natural levels." Pollution is "the addition of any substance or form of energy (e.g., heat, sound, radioactivity) to the environment at a rate faster than the environment can accommodate it by dispersion, breakdown, recycling, or storage in some harmless form". "Pollution is a special case of habitat destruction; it is chemical destruction rather than the more obvious physical destruction. Pollution occurs in all habitats—land, sea, and fresh water—and in the atmosphere."

"Much of what we have come to call pollution is in reality the nonrecoverable matter resources and waste heat." "Any use of natural resources at a rate higher than nature's capacity to restore itself can result in pollution of air, water, and land."

"Pollution is habitat contamination". Perhaps the overriding theme of these definitions is the ability of the environment to absorb and adapt to changes brought about by human activities.

In one word, environmental pollution takes place when the environment *cannot* process and neutralize harmful by-products of human activities (for example, poisonous gas emissions) *in due course* without any structural or functional damage to its system.

In fact, "the due course" itself may last many years during which the nature will attempt to decompose the pollutants; in one of the worst cases – that of radioactive pollutants – it may take as long as thousands of years for the decomposition of such pollutants to be completed. Pollution occurs, on the one hand, because the natural environment does not know how to decompose the *un*naturally generated elements (i.e., anthropogenic pollutants), and, on the other, there is a lack of knowledge on the part of humans on how to decompose these pollutants artificially.

Why does Pollution Matter?

It matters first and foremost because it has negative impacts on crucial environmental services such as provision of clean air and clean water (and many others) without which life on Earth as we know it would not exist.

INTRODUCTION TO ENVIRONMENTAL POLLUTION

Although pollution had been known to exist for a very long time (at least since people started using fire thousands of years ago), it had seen the growth of truly global proportions only since the onset of the industrial revolution during the 19th century. The industrial revolution brought with it technological progress such as discovery of oil and its virtually universal use throughout different industries. Technological progress facilitated by super efficiency of capitalist business practices (division of labour – cheaper production costs – overproduction – overconsumption – overpollution) had probably become one of the main causes of serious deterioration of natural resources.

At the same time, of course, development of natural sciences led to the better understanding of negative effects produced by pollution on the environment. Environmental pollution is a problem both in developed and developing countries. Factors such as population growth and urbanization invariably place greater demands on the planet and stretch the use of natural resources to the maximum.

It has been argued that the carrying capacity of Earth is significantly smaller than the demands placed on it by large numbers of human populations.

And overuse of natural resources often results in nature's degradation. It's interesting to note that natural resources had been stored virtually untouched in the Earth for millions of years.

But since the start of the industrial revolution vast amounts of these resources had been exploited within a period of just a couple of hundred of years at unimaginable rates, with all the waste from this exploitation going straight in to the environment (air, water, land) and seriously damaging its natural processes.

TYPES OF ENVIRONMENTAL POLLUTION

There are three major types of environmental pollution:

- Air pollution
- Water pollution
- Soil pollution (contamination).

Some of the most important air pollutants are sulfur dioxide, nitrogen dioxide, carbon monoxide, ozone, volatile organic compounds (VOCs) and airborne particles, with radioactive pollutants probably among the most destructive ones (specifically when produced by nuclear explosions). Water pollutants include insecticides and herbicides, food processing waste, pollutants from livestock operations, volatile organic compounds (VOCs), heavy metals, chemical waste and others. Some soil pollutants are: hydrocarbons, solvents and heavy metals.

SOURCES OF ENVIRONMENTAL POLLUTION

Fossil Fuel Sources of Environmental Pollution

In modern industrialized societies, *fossil fuels* (oil, gas, coal) transcended virtually all imaginable barriers and firmly established themselves in our everyday lives. Not only do we use fossil fuels for our obvious everyday needs (such as filling a car), as well as in the power-generating industry, they (specifically oil) are also present in such products as all sorts of plastics, solvents, detergents, asphalt, lubricating oils, a wide range of chemicals for industrial use, etc. Combustion of fossil fuels produces extremely high levels of *air pollution* and is widely recognized as one of the most important "target" areas for reduction and control of environmental pollution.

Fossil fuels also contribute to *soil contamination* and *water pollution*. For example, when oil is transported from the point of its production to further destinations by pipelines, an oil leak from the pipeline may occur and pollute soil and subsequently groundwater. When oil is transported by tankers by ocean, an oil spill may occur and pollute ocean water. Of course, there are other natural resources whose exploitation is a cause of serious pollution; for example, the use of uranium for nuclear power generation produces extremely dangerous waste that would take thousands of years to neutralize. But there is no reasonable doubt that fossil fuels are among the most serious sources of environmental pollution. Power-generating plants and transport are probably the biggest sources of fossil fuel pollution. Common sources of fossil fuel pollution are:

Industry:

- Power-generating plants
- Petroleum refineries
- Petrochemical plants
- Production and distribution of fossil fuels
- Other manufacturing facilities.

Transport:

- Road transport (motor vehicles)
- Shipping industry
- Aircraft.

Fossil fuel combustion is also a major source of carbon dioxide (CO_2) emissions and perhaps the most important cause of global warming. Learn more about the causes and effects of global warming here.

OTHER (NON-FOSSIL FUEL) SOURCES OF ENVIRONMENTAL POLLUTION

Among other pollution sources, *agriculture* (livestock farming) is worth mentioning as the largest generator of ammonia emissions resulting in *air pollution*. Chemicals such as pesticides and fertilizers are also widely used in agriculture, which may lead *water pollution* and *soil contamination* as well.

Trading activities may be another source of pollution. For example, it's been recently noted that packaging of products sold in supermarkets and other retail outlets is far too excessive and generates large quantities of solid waste that ends up either in landfills or municipal incinerators leading to *soil contamination* and *air pollution*.

Residential sector is another significant source of pollution generating solid municipal waste that may end up in landfills or incinerators leading to *soil contamination* and *air pollution*. How can we control environmental pollution? It's clear that fossil fuels are among *the* biggest sources of pollution. We need to find alternative renewable sources of energy which can replace fossil fuels in the future.

Green investment provides a great platform to explore and develop new and clean sources of energy such as solar electricity. Building your own solar panels and using diy solar energy systems to meet at least part of your home electricity needs is another emerging opportunity for diy enthusiasts. This can really make a positive difference to the environment and reduce current pollution levels.

ENVIRONMENTAL POLLUTION EFFECTS ON HUMANS, OTHER ANIMALS & PLANTS

General Environmental Pollution Effects

Miguel A. Santos notes that a very important aspect of the effect of pollution is its dose (or concentration) required to cause environmental damage. He defines *pollution response* as "the change in the effect of a pollutant in response to a change in its concentration".

In this respect, he identifies 3 different types of response evoked by the environment to different pollution concentrations:

- Linear effect
- Greater-than-linear effect
- Threshold effect.

In *the linear effect*, environmental damage increases linearly with pollution concentrations. In other words, " the total damage or risk is directly proportional to the accumulated exposure".

This effect occurs with radioactive substances as well as mercury, lead, cadmium and asbestos. In *the greater-than-linear effect*, environmental damage increases with an increase in pollution concentrations but at a decreasing rate. This means that, as pollution concentrations continue to increase the environmental damage will continue to decrease.

This is the case with thermal pollution. In *the threshold effect*, pollution produces no effect until a certain threshold in pollution concentrations is achieved. In other words, "so long as a given threshold is not exceeded, the damage from pollution would be completely repaired as quickly as it is produced". This effect is found with biodegradable pollutants. It is also important to mention synergistic effects of pollutants on the environment. While interacting with each other, pollutants can produce greater impacts than when acting individually. A good example of that is a synergy between asbestos exposure and smoking in causing lung cancer. There is no doubt that excessive levels of pollution are causing a lot of damage to human & animal health, plants & trees (including tropical rainforests) as well as the wider environment. All types of environmental pollution – air, water and soil pollution – have an impact on the living environment. The effects in living organisms may range from mild discomfort to serious diseases such as cancer to physical deformities (for example, extra or missing limbs in frogs). Experts admit that environmental pollution effects are quite often underestimated and that more research is needed to understand the connections between pollution and its effects on all life forms.

ENVIRONMENTAL POLLUTION AND TROPICAL RAINFORESTS

Pollution invariably disrupts the environmental services provided by tropical rainforests. Among other things, the rainforests serve as:

- Regulators of global and local climate patterns
- Storages of biodiversity
- Guarantors of local environmental stability
- Providers of habitats for rainforest animals and forest people
- Storages of vast economic resources for human use.

So these vital services are seriously disrupted by air, water and soil pollution which all cause damage to the rainforests and people living in or around them.

The big relationship between environmental pollution and tropical rainforests appears to be three-fold:

- Pollution is brought upon the rainforests, i.e. the rainforests act as *pollution recipients;*
- Certain uses of the rainforests by people cause pollution to the wider environment, i.e. the rainforests act as *pollution sources* facilitated by human actions; *and*
- Rainforests "soak up" certain types of environmental pollution and neutralize it, i.e. they act as *pollution sinks.*

As an example of *point 1,* waste dumped into rainforests can be generated by industrial or agricultural activities. For instance, production of oil by Texaco in Ecuador rainforest led to this company dumping millions of gallons of toxic waste into pristine rainforests that acted as pollution recipients. *Point 2* is well illustrated by the process of rainforest destruction which leads to pollution of the forests and wider environment. Activities such as clearance of rainforests for growing "cash" crops, cause the release of carbon dioxide into the atmosphere and contribute to global warming. Here rainforests act as pollution sources. In *point 3,* rainforests act in their unique capacity as pollution sinks – they remove carbon dioxide from the atmosphere and thus are capable of neutralizing pollution.

TROPICAL RAINFORESTS AS POLLUTION SINKS & SOURCES

A *sink* is defined as "the process, activity or mechanism that removes a greenhouse gas, aerosol, or precursor of a greenhouse gas from the atmosphere". Another known type of greenhouse gas sinks are oceans. As we know, greenhouse gases such as carbon dioxide (CO2) are serious pollutants which contribute to climate change on Earth, specifically global warming. We have seen in point 3 above that tropical rainforests act as environmental pollution sinks.

For example, cars are among the biggest producers of CO2 emissions in the world. Car emissions produced anywhere in the world may be absorbed and removed from the atmosphere by rainforests located in tropical countries. By absorbing these emissions, rainforests help relieve the global warming pressure. So from this point of view, tropical rainforests are uniquely placed to help regulate and stabilize global climate patterns.

In other words, by acting as pollution sinks rainforests provide a common *global* good for the rest of humanity. On the other hand, if rainforests are destroyed they will also act as major sources of global warming pollution. This happens because rainforests are huge storages of greenhouse gases. And as a result of deforestation these greenhouse gases will be released back into the atmosphere and thus will further contribute to global warming.

Deforestation is a very serious problem - it has recently become one of the main causes of global warming. Global warming may, in its turn, have a negative impact on the rainforests. For example, warmer temperatures may cause an accelerated growth of pests which destroy the trees. Learn more about it in our article on effects of global warming.

So it is in the interest of all nations to have the rainforests intact and healthy. But rainforests have a limited capacity for acting as pollution sinks. There is only so much environmental pollution that the rainforests can soak in. We need to find ways to reduce pollution and bring it under tight control. From this point of view, green investment offers great opportunities for developing clean technologies for the future.

SOURCES AND EFFECTS OF TROPICAL RAINFOREST POLLUTION

Rainforest pollution has negative impacts on people, animals, trees & plants as well as the wider environment. Sources of environmental pollution affecting rainforest ecosystems can be both *natural* and *anthropogenic* (human-caused).

Local Anthropogenic Sources and Effects of Rainforest Pollution

Agriculture is a well known source of environmental pollution, specifically causing ammonia emissions from livestock farming and animal waste. Agricultural activities in rainforest areas are no exception to that. But in addition to ammonia emissions, rainforest clearance by fire for agricultural purposes (especially cattle ranching) is also a source of harmful carbon dioxide emissions which contribute to global pollution and global warming.

Farmer settlements:

- Use of "slash and burn" method to clear land for agriculture; the burning process releases carbon dioxide into the atmosphere contributing to air pollution and global warming
- Rainforest soils are poor in nutrients; this leads settlers to continuously seek new land and clear even more rainforest by burning
- Where there are human settlements, there will always be human-generated waste left behind causing further rainforest pollution.

Cattle Ranching:

- Forest clearance by fire for cattle ranching releases more carbon dioxide into the atmosphere further contributing to environmental pollution and global warming
- Overgrazing causes severe soil degradation (2)
- Contributes to ammonia pollution.

"Cash" crops growing:

- Widespread use of p esticides (for example, for soybean or sugarcane cultivation) (3) may lead to major pollution of water and soil
- On the positive side, nitrogen pollution coming from the use of nitrogen fertilizers can actually boost the growth of plants in tropical rainforests.

Industry

It's not difficult to imagine that industrial sources are major contributors to rainforest pollution.

Some of the most important industrial activities causing pollution are summarized below:

Logging / Timber extraction

- Waste wood chips ("slash") from logging left behind pose a danger in case of forest fires
- Use of heavy machinery may lead to soil compaction as well as leaking of oils used for managing such machinery
- Removal of trees can cause nutrient depletion [for rainforest ecosystems]

- Removal of trees leads to the release of carbon dioxide into the atmosphere contributing to air pollution and global warming
- Logging companies will often slash and burn large areas of forest just to get access to one commercially valuable tree; such [mindless] forest burning will release more carbon dioxide into the atmosphere
- Logging leads to deforestation / rainforest destruction and the loss of potential pollution sinks and their ability to remove carbon dioxide from the atmosphere
- Land cleared by logging helps new farming settlers make their way into the rainforests and cause more deforestation / pollution.
- Logging also destroys rainforest animals' habitat; many of rainforest animals are now endangered due to habitat loss and environmental pollution.

Mining

- Mercury used in extracting the gold during the "Gold Rush" of the 1980s in Brazil's Amazon, polluted waterways causing the fish to be inedible
- During mining activity, blasting releases particles and noxious fumes; the winds blow across open-cast mines, waste heaps and dumps of toxic mined products like asbestos, spreading the pollution
- Dust from mining can also negatively affect agricultural fields.

Hydroelectric Dams

Hydropower is a major source of energy supplies in developing rainforest countries. For example, around 83% of total power production in Brazil in 2004 came from this source.

Construction of dams used to generate hydroelectric power is a serious cause of rainforest pollution.

The effects of dam construction on rainforests are as follows :

- Large areas of the rainforest are flooded, thousands of people living in the vicinity are uprooted and moved
- Rates of waterborne diseases increase
- Dams trap silt that contains valuable nutrients for aquatic life, thus damaging downstream ecosystems
- Reduced levels of silt cause coastal soil erosion
- Water pollution that damages and kills aquatic life.

EXAMPLE: – the Balbina dam, Brazil:

- 1,400 sq. miles of rainforest flooded, with trees rotting in water
- Water pollution:

o Water acidified, undrinkable with heavy metal pollution

o Large number of fish have died.

Oil Production

Oil production and its related activities are infamous around the world for their capacity to bring pollution upon virtually any life-hosting environment that comes in contact with them, including water, soil and air.

Oil-related activities have been known to pollute vulnerable and highly valuable rainforest ecosystems as well. An excellent example of that is the above mentioned case of Ecuador rainforest pollution by Texaco oil operations.

Below we discuss some effects of oil production on people and rainforest ecosystems:

- Oil spills in coastal areas as well as on land due to oil production and transportation cause water pollution, soil pollution and air pollution. Some of the effects are as follows:
  - Cancers and tumors, skin and breathing problems, gastrointestinal disorders in humans (13) [and possibly animals]
  - Physical deformities in newborns
  - Destruction of aquatic life and aquatic ecosystems in general.

ENVIRONMENTAL POLICY IN THE SERVICE OF HUMAN RIGHTS

There can be no question, however, that the safeguarding of human rights is more urgent than the achievement of fairer distribution – especially at the level of the world. Survival comes before a better life. The unconditionality of human rights may therefore serve as the basis for the setting of priorities: the realization of fundamental rights must take precedence over all other activities, including the realization of non-fundamental rights. Applied to ecological subsistence rights, this means that the right to a living must take precedence over the non-fundamental resource needs of other agents. Subsistence needs come before luxury needs. This formula indicates the basic duty that a recognition of subsistence rights implies for national and international institutions. Since everything depends on progress in closing the power scissors, we might speak here of a dual strategy: to increase the room for manoeuvre available to the poor, and to limit the power of the well to do.

For the poor to achieve greater room for manoeuvre, local community rights over resources must be recognized and strengthened in cases of conflict. After all, pasture and forest, fields and seeds, fresh water and clean air are valuable sources of nourishment, health, materials and medicine. This is why a policy that protects livelihood rights overlaps with the interest in environmental conservation. Since intact ecosystems mean that the poor are less vulnerable, conservation of nature and the environment is the core of any serious policy to overcome poverty. And, conversely, since effective rights for local inhabitants are the best guarantee that the resources of the poor will not be easily diverted to the rich, a policy of protecting people's right to a decent living is a central plank of natural and environmental conservation. Ecology and subsistence rights are thus very closely intertwined.

The aim of limiting the power of the well-off may be based upon elementary principles of fairness. We should not think here of redistribution between rich and poor, but rather of what we might call the minimum principle of justice: that is, all national and international regulations should be crafted in such a way that they do not worsen the lot of the most disadvantaged. It appears to be a modest principle but is actually quite tough. For the cross-border economic and ecological consequences of production processes, foreign investment, protectionist measures or financial transactions are so enormous that such a principle would bring about a major change of priorities in economics and politics. Both investment decisions and multilateral policy negotiations are characterized by efforts to gain an advantage over rivals – without much heed to the costs for the most disadvantaged sections of the population, who usually have no place at all at the decision-making table.

Finally, a transition to sustainability in the more affluent economies, in both the North and South, is a necessary condition for the safeguarding of the subsistence rights of those whose livelihood depends on direct access to nature. In the short term, more efficient fuel use and agricultural production can ease the pressure on life-serving ecosystems and local communities; greater bargaining-power can also enable local communities to get more compensation and a larger share of the profits. In the longer term, however, the conflicts over environmental human rights can only become sharper if the global class of high consumers continues to maintain its demand for natural resources. Only if demand for oil falls will it no longer be worth launching drillings in the primeval forest. Only if the thirst of agriculture and industry abates will enough ground water remain to supply the village wells. Only if the burning of fossil fuels is restricted will insidious climate change no longer threaten the existential rights of the poor. In sum, resource-light models of production and consumption in affluent economies are called for by a respect for human rights. For the statistical fact that a minority of prosperous countries overburden the global environment is now becoming a palpable reality as it leads to the degradation of other societies. Far from conserving only whales and yellow water-lilies, environmental policy is one way of ensuring that our finite world remains hospitable to a growing number of people. It is in this sense that the Narmada Valley demonstrations have been pointers for a future where environmental human rights are protected against the resource hunger of high consumers in North and South.

EXPLOITATION OF TRADITIONAL COMMUNITIES

The intersection of trade, environment and human rights reflects the collision of powerful and competing forces. The strategic question for human rights activists is how those colliding forces can be brought into at least a rough balance in which the power of the market is used to advance our humanity rather than weaken it. There is little question that uncontrolled market dynamics have caused serious problems in the spheres of human rights abuse and environmental protection. Lofty rhetoric about the benefits of economic development too often masks the unfair allocation of the costs and benefits of development. A former official of the World Bank once admitted during a seminar on cost-benefit analysis in which I participated that in many of the Bank's projects the benefits went to the more well off people in developing countries while the costs were disproportionately borne by the already disadvantaged.

Globalization and Industrial Shrimp Aquaculture

This chapter explores the collision of competing systems and values in the specific context of the human rights consequences of large scale industrial shrimp aquaculture—a form of lucrative international commodity trade that has exploded in scale over the past twenty years and is consuming the coastlines of many developing countries. In addition to undermining the coastal ecosystems, industrial aquaculture is ruining the lives and harming the livelihoods of many traditional users of coastal resources.

While aquaculture has been practiced for many centuries, industrial scale shrimp aquaculture is of very recent origin in most instances starting up no more than two decades ago. Shrimp aquaculture offers a paradigm case of the inherent conflict between the processes of a globalized economy and the rights and interests of local people. Shrimp aquaculture is one of the points at which the demands of a market economy, the dominance of capital, the need (or desire) for export-driven foreign exchange earnings and the rhetoric of globalization clash most intensely with the dynamics and fundamental values of more traditional user economies and smaller scale fishing and agriculture activities.

Industrial shrimp aquaculture was created in tropical developing countries as a high value cash crop to feed restaurants and supermarkets in the United States, Europe and Japan. Shrimp is a key part of the menu and an important determinant of the profitability of restaurants. It also sustains the seafood sections of supermarkets in the developed countries of the North, representing billions of U.S. dollars in annual sales. The result of the profit to investors that results from the high demand for shrimp from the restaurants and supermarkets has been uncontrolled growth of extensive aquaculture farms on the coasts in many tropical developing countries. The typical pattern is that domestic and foreign investors with few or no ties to local communities exploit the areas and undermine the base of natural productivity on which the traditional residents depend. This has destroyed the livelihoods and rights of many traditional users of the coasts and smaller communities whose members have had no voice in the development.

The harms produced by large-scale shrimp aquaculture are in part an effect of the sheer scale of operations on fragile coasts and ecosystems that were never designed to sustain such heavy burdens. In Thailand and other shrimp producing countries, miles of shrimp ponds have been placed end-to-end and side-to-side in ways that alter the coastlines and block traditional users' access to coastal resources. Before the extensive aquaculture farms came into place the coasts were typically filled with mangrove forests that served the needs of local users. Now the ponds change and deplete the ecosystem, pollute the waters, destroy the mangrove forests and their productive capacity, reduce the quality of fishing, introduce salt into freshwater supplies, and harm local agriculture. They also lead to fences, guard towers, human rights violations, and land-grabs by the shrimp farming industry.

But equally significant with the issue of excessive scale of operations in fragile ecosystems has been the overall irrationality and lack of effective government planning and monitoring that has characterized the expansion of shrimp aquaculture. Its advocates among international and bilateral development institutions have seen aquaculture as a key element of an economic revolution that would bring jobs and opportunity to poorer developing countries. In most instances the intentions of the international institutions were benign. They only saw shrimp aquaculture as a means to improve the economic development of poorer countries. But the international development institutions, as has far too often been the case, chose to remain ignorant about the full range of effects created by their development projects and about who gained and lost in the process.

But many officials in national governments as well as individuals who invested in shrimp aquaculture had no benign social concerns. They have seen only the opportunity for quick profits through the exploitation of coastal areas that previously had been seen as too difficult or expensive to develop. Until the aggressive expansion of shrimp aquaculture the remote coastal areas had been left alone by investors and government officials because they perceived no sufficiently profitable way they could use the areas. Shrimp aquaculture changed this perspective. The problem is that the coasts were already being used as a productive commons system by existing residents of the areas that have now been invaded by the shrimp farms.

The first step taken by many of the shrimp producers was to gain user rights to those coastal areas from friendly governments that were already close to the new investors. Very often government officials took an equity share in the shrimp operations. This created a financial incentive for government officials to grant user or ownership rights to the new shrimp producers and investors at the expense of the local communities and traditional users. Given these developments and incentives, producers were allowed to operate in a poorly regulated context with virtually no oversight or protection of traditional users and local communities. This has caused unnecessary harm to communities and environments in many nations where the industrial shrimp producers have located their farms. Along with these disruptions have been numerous instances of human rights violations.

Human Rights Violations

A recent report from Bangladesh shows the intensity of the conflict in many countries. A news report described how people desiring to develop shrimp farms in a contested area resorted to arson and murder to drive out the residents. The report from The Daily Star , dated October 4, 2000, stated that: "About 50 makeshift houses were torched at Kaliganj-Lebukhali under Baniapara thana here Monday by miscreants, allegedly led by OC [Officer in Charge] of a local police station. More then 60 people including women and children were hurt during the incident. The miscreants also hurled hand made bombs at the villagers. Villagers, who tried to resist, suffered serious burns while about 350 families living in the makeshift houses on the khas land [land leased from the government] fled fearing attack. The motive behind the attack was to grab the khas land for shrimp culture, sources said.

I recently traveled to Bangladesh and spent part of the trip visiting a memorial dedicated to a Bangladeshi woman, Korunamoyee Sardar, who had been a leading protester against the encroachment of shrimp farms into the area where she lived. She was murdered and beheaded, and her head impaled on a tree limb in an effort by powerful interests to intimidate others who might oppose their desires. But her martyrdom only symbolizes the widespread cancer that shrimp aquaculture has brought to numerous developing countries. Our investigation in Bangladesh revealed that local police ignore violations, that they assist the "Shrimp Lords" in intimidating and filing false charges against local villagers; that murders, rapes and beatings by thugs working for the Shrimp Lords are common, and that the police and judicial system look the other way.

In Thailand many local authorities fear to enforce aquaculture laws against many producers because of the threat of violent reaction. The fears are not hypothetical. It was recently reported that, "Siripote Cheechang, a member of the tambon Pa Khlok Administrative Organization who led local people against mangrove forest encroachment, was severely injured when hit twice by a car. Police believe the incident was attempted murder plotted by some forest encroachers." Even more recently some shrimp farmers in Thailand appear to have moved from attempted murder to murder. Jurin Ratchapol, an activist leading the effort against illegal shrimp farming was shot to death. The report in The Bangkok Post reports that: "His family said he had received death threats from workers at the Watchara prawn farm who wanted him to stop his conservation activities. Jurin's 41-year-old widow, Ladda, said his murder could not force her to give up the fight against over-exploitation of natural resources. Like other local villagers, the Ratchapols earn their living from small-scale fishing and gathering of forest products."

The Clash over Land and Water Rights

It should not be surprising that many of the violations of local peoples' traditions and rights involve disputes over the preservation and use of land and water. Shrimp aquaculture has intruded into mangrove ecosystems and other resource systems on which many people have traditionally depended for their livelihood. Much of the conflict over shrimp aquaculture comes down to the protection, allocation, and management of land and water rights. Land and water rights are points of severe disagreement and tension in the siting of shrimp aquaculture. In many shrimp producing nations the coastal zones that are the primary sites of shrimp farms are commons areas freely used by residents for many generations. These areas are often owned by the State with the government having the authority to grant concessions or even sell the land and use rights to private owners. The desire to protect the traditional users of land and coastal zones that were not privately owned lies at the core of the Code of Conduct for Responsible Fisheries. Yet the CCRF contemplates a shift to private land rights as an important part of the expansion of shrimp aquaculture.

A significant part of community and environmental opposition to the growth of shrimp aquaculture has stemmed from the widespread devastation of mangrove forests. Most shrimp producers and government development and finance ministries have tended to view mangroves and other wetlands ecosystems as nothing more than trees, salt flats and seasonal lagoons that are unproductive areas that are obstacles to "real" economic development. Those who live in or near the mangroves and similar "commons" systems see them very differently. To those people such areas are the critical element in a highly productive ecosystem on which their livelihoods and culture depend.

Globalization and its emphasis on the market economy demand private property rights recognized by law. In a private market economy a person seeking to establish a shrimp aquaculture farm or needing additional capital for expansion or improved operations needs to acquire rights to the necessary land. For those raised on the sanctity of western market economics this fact seems inevitable and even desirable. But for more community-oriented and smaller-scale systems the approach is destructive.

Acquiring land and user rights will often be a matter of buying the land outright, or acquiring land through a lease or government concession. No rational private sector economic actor will invest significant private funds in a venture over which it lacks long-term control. Any rational shrimp farmer would be a fool to make long-term investments in a shrimp farm operation in a situation where the farmer lacks clear title to the land or holds only a short-term concession from the government. Even if the concession is renewable every five or ten years investors can't take the chance and make the kind of infrastructure investments required to become a fully responsible shrimp producer. Without long-term ownership or use rights a farmer will not make capital investments in the operation beyond what is reasonably necessary to maximize short-term returns.

Globalization and Its Impact on the Less Powerful

Globalization has become a code word for the massive and penetrating expansion of market-based economic activity throughout the world. Thomas Friedman has described globalization as involving "the inexorable integration of markets, nation-states and technologies to a degree never witnessed before in a way that is enabling individuals, corporations and nation-states to reach around the world farther, faster, deeper and cheaper than ever before, and in a way that is also producing a powerful backlash from those brutalized or left behind by this new system."

The phenomenon of globalization has resulted in sharp increases in the volume and value of internationally traded products, and the expansion of regional trade and economic alliances such as the European Union, NAFTA, GATT and the World Trade Organization (WTO), ASEAN, and Latin America's Mercosur. More arrangements of this type are on the way. As this transformation and realignment of world economic activity and supporting institutions has occurred there has been a shift toward greater privatization of economic power and the creation of various versions of free-market capitalism. An important result has been the dilution of formal governmental power over business enterprises, especially those enterprises with a global reach. Accompanying and supporting this change has also been the enormously increased scale and rate of mobility of private capital flows. A 1999 World Bank report related that developing countries received $155 billion in net direct foreign investment in 1998. This was more than 16 times the amount of net direct foreign investment seen in 1990. Net capital flows from government development assistance programs declined from slightly above $50 billion to slightly below $50 billion during the same period.

Human Rights, Land Use, and Market-Driven Globalization

Private property, or at least the legal right to claim exclusive long-term user rights for land and coastal areas, represents a fundamental challenge to traditional cultures. The problem is that private property is inherently exclusionary. To a market-oriented entrepreneur the resulting ability to have a monopoly right to extract the value from land is a desirable good. But to a person dependent on being able to harvest food and other products from a commons area and to protect against the effects of pollution of land, water and local fisheries, intensive development of private use rights can destroy an area's culture, community and economy that has been developed over generations in a careful balance with nature. The clash between the different systems can be managed only with great care. This requires effective regulation, laws and institutions and they simply do not exist in any meaningful combination.

The UN Food and Agriculture Organization's Code of Conduct for Responsible Fisheries provides a benchmark and frame of reference that in theory include protecting the environment, respecting the rights of local communities, and contributing to improvements in social justice while still managing to build sound economic activity. The reality of what exists in the field and how national governments and producers have failed to take responsible steps to implement the Code of Conduct has proved to be different. At present there are frequent proclamations of good intentions by the shrimp industry, by national governments, and international financial institutions but with virtually no serious follow through to implement actions to protect the lands and rights of local communities. The harmful impacts continue to occur at the same time that governments and industry issue public relations statements proclaiming their commitment to "sustainable" aquaculture that protects communities and traditional users. But those proclamations are public relations oriented window dressing, not reality.

The challenge of achieving responsible shrimp aquaculture that limits or avoids severe impacts on traditional communities and cultures, protects against environmental and ecosystem degradation, and does not countenance human rights violations can not be overstated. It is a fundamental political, economic and cultural clash between the haves and have-nots. One approach could be to responsibly implement the FAO's Code of Conduct for Responsible Fisheries. It is dedicated on paper to ensuring the improvement of the quality of life for poorer communities. It recognizes the importance of efforts to create investment models that ensure participation of local people in ways that result in enhanced training and the ability to share in the benefits of economic development. But this has not been done. One of the reasons is that many of the countries that produce farmed shrimp have rigidly stratified social systems in which there is no space or desire on the part of the most powerful to facilitate the upward mobility of "lower" classes.

Corruption of a Legal System as a Human Rights Violation

Human rights traditionally made a distinction between state and private action. The lines are becoming increasingly blurred as a great deal of "public" power is mirrored in the global scale and enormous power of the private sector to such a degree that it is appropriate to consider the more powerful corporate actors as wielding a form of quasi-state power in relation to which it is necessary to apply human rights responsibilities to those quasi-state actors. In any event, law is one of the fundamental areas through which state power is exercised and for which states should be held accountable for serious deviancy or failure to use state power to act appropriately or to protect citizens from abuses.

A serious issue of what constitutes a cognizable violation of human rights involves the corruption of a nation's legal system. When fair treatment can not be consistently guaranteed, the Rule of Law does not really exist—regardless of form. There are, of course, issues when a legal system performs badly for everyone across the board. At some point matters are below the human rights "radar". But when a legal system is controlled by corrupt interests for their own purposes without recourse for those harmed by its decisions, there has been a violation of human rights. Jorge Varela's description of the inadequacy of the Honduran legal system, its refusal to enforce numerous existing laws against shrimp producers who violate laws with impunity, and the lack of any prosecutions following the murders of fishermen near shrimp farms offers a powerful example. Similarly, the lack of any real penalties in Ecuador for violations by shrimp farmers engaged in illegal activity, the presence of corruption and threats addressed to regulatory officials, and the inability to convince judges to act against violators repeats the pattern found in Honduras, Bangladesh, India and elsewhere.

In a system where corruption is present the best farmers might well find their concession not renewed because someone else wants to take over their high quality shrimp farm. Investors in Bangladesh and India have told me that there is a danger in their countries of such behavior when powerful local figures decide to take over their operations, particularly for smaller-scale farmers with virtually no resort to the legal system.

Legal systems in many of the nations where shrimp aquaculture has grown most rapidly are either generally ineffective or skewed toward the interests of those with power and capital. Most nations are ill-equipped to create the complex, expensive and integrated management systems that would be required to satisfy the demands of market economy entrepreneurs and simultaneously protect the needs of existing communities or "commons-based" economies and cultures while preserving the quality and productive capacity of natural resources and ecosystems. Almost without exception, exclusionary private land rights are threats to more traditional and dispersed local communities of the kind found in the coastal zones most favored by shrimp farmers. The more intensive economies of scale that characterize export-driven industrial shrimp farming deplete or harm natural resources that are essential for the health of the local economies.

6

The Global Compact and Human Rights

REFRESHER ON THE GLOBAL COMPACT

The Global Compact was introduced in 1999 because many felt that globalization was not delivering all it had promised. In particular, its benefits were dramatically unevenly distributed. Although the primary responsibility for implementing international agreements on human rights, labour standards, the environment and anti-corruption lies with governments, business was seen as having an important role to play. In fact, one of the founding premises of the GC is that without active involvement of business, there is the danger that universal principles, including those embodied in international human rights instruments, will remain unimplemented.

The Global Compact emphasizes the business case for improved social and environmental performance and business' own operations and activities not the traditional CSR concept that is based on philanthropy. We believe that the largest impact that business can have is through conducting their business in a sustainable way that respects human rights and limits their environmental impact, treats their workers well and avoids corruption.

The Compact, itself, is both a set of universal principles and a network-based multi-stakeholder initiative with a change model for how to raise the level of business' social and environment performance. The Secretary-General derived the principles from international instruments that enjoy wide consensus like the UDHR, ILO Declaration, Rio Declaration, and UN Convention Against Corruption.

Businesses initiate their participation through a commitment by their CEO endorsed by their Board expressing their support for the principles and their intention to work towards their implementation within their own operations. This is a key element of the Compact's leadership model. Effective positive change requires a leadership commitment and it is our intention that change agents inside companies will use their CEO's expression of commitment to help them to more easily make necessary changes inside the company to improve its social and environmental performance. The Global Compact is not a seal of approval for or endorsement of the companies or other organizations participating in it. Nor is it a regulatory or compliance based model. The Global Compact Office has neither the mandate nor the resources for that. As a result, we encourage improved performance and positive change through the means available to us as an initiative inside the UN Secretariat: learning, dialogue, networks, initiatives, projects and promoting greater transparency around company social and environmental efforts. We also work with the financial incentive system within which companies operate to encourage the mainstream financial community to take into account companies' social and environmental performance in making their valuations.

Global Compact Office was established in July 2000 with less than 50 companies. Today, there are 2268 participating companies in 87 countries, dozens of civil society and labour organizations and more than 40 local networks engaged.

THE GLOBAL COMPACT'S APPROACH TO HUMAN RIGHTS

Two of the Global Compact's ten principles are devoted to human rights. However, the labour standards principles are also human rights related. The two human rights principles are positive and negative, that is, they emphasize doing good as well as not doing harm. Human rights is a major priority area of the UN Global Compact Office as it remains one of the newest and least familiar for many of the company participants. We've been doing a fair bit of work in this area and plan to continue doing so. We feel we are making an important contribution although much more clearly remains to be done.

Our main partner in all our human rights work has been and will remain the OHCHR. They are clearly the guardians of the two human rights principles. We also work with a number of other organizations in this area, including BLIHR and the Danish Institute for Human Rights.

Located within the UN Secretariat-the Secretary-General's office to be precise-the Global Compact Office must be as pragmatic and non political as possible. We focus on the practical "How" question, namely, how to implement human rights within a business's own operations and activities, especially on what businesses can do to eliminate their potentially negative human rights impact. Two recent publications that we have produced in this area are:

Raising the Bar; and Embedding HR in Business Practice. One of our major activities in this area has been fostering tool development and recommending proven tools to help with the implementation process. Rather than reinvent the wheel, we help disseminate information about existing tools with a proven track record, such as the Business & Human Rights Resource Centre's website, the Danish Institute Human Rights Compliance Assessment quick check tool, and the BLIHR matrix. We are also working on a human rights training package and, with partners, on a business guide to human rights impact assessment and, with BLIHR, on guidance for managers on how to implement human rights. One of our most significant contributions to the area continues to be our efforts to build consensus among companies, other societal actors and even governments that human rights are an important business issue. Our leadership model is key here. The process for initiating involvement (CEO commitment with endorsement from the Board) means that many boardroom and/or senior executive discussions about human rights have taken place because of the Global Compact. Moreover, 2268 companies are on record as having expressed support for and intention to implement human rights. Importantly, these are companies in 87 different countries, helping to demonstrate the universality of human rights. Governments have also collectively and individually recognized the contribution of the Global Compact in a number of contexts, most recently in the 2005 World Summit Outcome document.

We are pleased to report that much learning is going on at the local level. We know that some Global Compact local networks have used our human rights publication as the basis for running their own learning activities on human rights. Many Global Compact local networks have also invited local chapters of human rights NGOs to come and speak with them to help them learn about human rights in a business context. In some countries where governments are not very supportive of human rights, business has been taking a keen interest and are slowly helping their governments to see that human rights are important.

We have also contributed to the business and human rights debate through helping to develop and refine its lexicon. As already mentioned, we helped popularize the concepts of "sphere of influence" and "complicity" as key terms that have helped focus and advance the business and human rights debate. We also collect and disseminate case studies of the real dilemmas that companies are facing and provide opportunities for companies and others to come together to work on solutions. E.g. Supply chain, business in conflict zones (Enabling economies of peace), managing diversity and non-discrimination, and contributing to OHCHR's consultation process last year. As the UN, we can bring to the table audiences that others sometimes can't.

Is our approach having an impact? McKinsey & Company's impact assessment in 2004 found that the Compact has existential power and has had significant incremental impact in the sense of helping change to happen faster and more easily-which is our modest aim. Among the other findings were that half of all participating companies reported having changed their policies with respect to the principles. Many companies from OECD countries have signed on to Global Compact to learn about how to improve their human rights performance. For nearly two-thirds of the companies from developing countries, the Global Compact is the first corporate citizenship initiative in which they have engaged. Many do so to learn how to improve their social and environmental performance and thus enhance their ability to enter into supplier relationships with larger global firms.

There is no doubt that much more remains to be done. One of our most exciting upcoming challenges is that we're bringing our multi-stakeholder approach to China with a big Summit there later this year.

ROLE OF THE GLOBAL COMPACT

A word or two is also warranted on the role of the Global Compact in the context of calls for greater legal accountability for business. The first point to note is that the Global Compact is a voluntary initiative without a mandate or resources to monitor compliance. Nevertheless, protecting the integrity of the UN and of the initiative is a legitimate concern. We therefore introduced a stricter logo protection policy and new integrity measures covering misuse of the GC logo and a complaint procedure for systematic and egregious abuse of the GC's overall aim and principles. The policy goes as far as UN legal office will allow given our status as voluntary initiative of UN Secretariat.

We also introduced a procedural requirement that all company participants communicate to their stakeholders on a regular basis their progress in implementing all ten principles. Companies not doing so for two years in a row are designated "inactive" on our website until they start communicating.

This has already started to show results with media calling attention to "inactive" companies, prompting them to start communicating. Most importantly of all, we cannot emphasize enough that the Global Compact's voluntary initiative is not intended to be a total solution. It is positioned as complementary to and not a substitute for other approaches, including for international law. Human rights issues are complex and a multitude of approaches is clearly needed.

THE INTERNET AND HUMAN RIGHTS: AN OVERVIEW

The Internet is a unique communications medium. Like no other medium before, it allows individuals to express their ideas and opinions directly to a world audience and easily to each other, while allowing access to many more ideas, opinions and information than previous media have allowed. Consequently, there is a vital connection between the Internet and human rights.

Through the Internet, citizens from the most repressive regimes are able to find information about matters concerning their own governments and their human rights records that no newspaper may dare print, while denouncing the conditions under which they live, for the world to hear. The Internet allows an intimate look at other countries, other people and other cultures that few before were ever able to attain. This power to give and receive information, so central to any conception of democracy, can be truly achieved on the Internet, as nowhere before.

Further, through the use of encryption technology, citizens can have instantaneous communications with individuals all over the world that are much more resistant to government and private surveillance.

On the Internet, citizens are not mere consumers of content but also creators of content. This fundamental shift in power has created a possibility for every individual to be a publisher. Consequently, the content on the Internet is as diverse as human thought. Individuals and communities have been using the new-found freedom online to link, interact and work collectively in this global work space.

The effect of access to and use of this global interactive medium has been to promote and defend civil and political rights worldwide. This unprecedented power, however, can be very threatening to repressive regimes. The experiences of communities in different countries indicates that few things could be more threatening to authoritarian regimes than access and use of a medium that knows no boundaries and is very hard to control. While traditional methods of censorship-embargoing newspapers and closing down presses-do not work on the Internet, the online censoring techniques that these regimes attempt can be just as destructive. While the Internet is technologically resistant to government control, it is not immune from such control. Indeed, some countries have been quite sophisticated in exploiting the control and surveillance potential to great effect, at least in the short run. Just because the younger generation may know how to "hack" through proxy servers to avoid censorship does not mean that the youth are safe. These actions should be understood in the technological context--that such hacking is probably obvious to government system administrators, and may make people vulnerable to being identified and prosecuted. Meanwhile, nations around the world are seeking to exploit the surveillance potential of this new medium, including by asserting control over the design and development of communications networks to maximize their surveillance capabilities.

For these reasons, governments and regional and international bodies should enact enforceable free expression guarantees, promote widespread access to and use of strong cryptographic tools that will enable real security, and provide legal protections of privacy, including controls on government surveillance.

The Internet and Human Rights Work

The Internet is one of the best means for communicating on human rights, because it is inexpensive and global. E-mail makes point-to-point communication between human rights workers and among NGOs (non-governmental organizations) cheap and easy, and allows for better coordination of actions. Furthermore, the Internet has the potential of reaching global audiences, including those most in need of such information. The Internet is important to those working for human rights, as it can provide a secure means of communicating between and coordinating the work of human rights groups. Consequently, human rights activists were among the first to make use of the Internet to--

- Coordinate actions and make contacts. Perhaps the most important application of the Internet in human rights is as a great tool for NGOs and activists to be in touch with each other, to share information privately, and to coordinate actions.
- Expose human rights violations and let people know about them. Many human rights organizations throughout the world have instituted e-mail lists to propagate their press releases, alerts and denunciations vis a vis human rights violations. Many of these groups also use Usenet to post their information, or it is posted there by others, and many groups have instituted web pages to make information available.
- Solicit action. Many organizations now use e-mail to distribute action alerts calling for letters to be written on behalf of specific issues.
- Gather information. The amount of human rights information online is extraordinary-beginning with the University of Minnesota web site, and now with the UNHCHR (United Nations High Commissioner on Human Rights) site online, it's much easier to do human rights research.

By enabling early access to information, immediate dissemination of calls for campaigns, and the organization of wide international pressure, the Internet greatly increases the lobbying capacities of human rights groups.

It is vitally important, therefore, to promote freedom of expression and privacy as central elements of the development of the information society.

Internet Freedom in the US

Unfortunately, the US government's domestic policy has not been fully supportive of human rights on the Internet. The US Congress has twice enacted censorship legislation attempting to control content on the Internet. Under our strong constitutional protection of free speech, these laws have been ruled unconstitutional by the courts. Yet Members of Congress continue to press other restrictive measures, and proposals are pending to make Internet Service Providers (ISPs) liable for removing allegedly illegal or improper content. On the privacy side, the Communications Assistance for Law Enforcement Act of 1994 (CALEA) requires that telecommunications systems be designed to accommodate government surveillance, and that law has been perverted by the Department of Justice and the Federal Communications Commission to not merely preserve but to enhance government wiretap capabilities. Only recently has the US government begun to turn away from its efforts to control the spread of encryption, which is crucial to security and privacy on the Internet. Still, officials in the Justice Department are trying to preserve their access to decryption keys, and the surveillance debate in the US is not over. Meanwhile, the US government refuses to publicly discuss its international eavesdropping activities, amid growing domestic and international concern over the nature of surveillance in a digital world and, in particular, the US government's "Echelon" surveillance system.

Internet Access Is Important to the Promotion of Human Rights

The most often disregarded impediment to the achievement of the Internet's democratic potential is the problem of access. Accessing the Internet requires either the possession of a computer and payment to an Internet Service Provider or access to a facility where one can connect. This means that, generally, only the affluent will have access. Governments can control freedom of expression of sections of the population by hindering their access to the Internet. This may be done directly by imposing high taxes on computer products and supporting telephone monopolies that keep phone rates high and thus discourage the use of the Internet, or indirectly by failing to support community centres where people may have low-cost access to the Internet.

However, augmented Net access without civil liberties guarantees ? in technology and in law ? may not benefit individuals and may be coupled with government control and extended surveillance capability.

Freedom of Expression

The Universal Declaration, the European Convention and other international human rights agreements enshrine the rights to freedom of expression and access to information. These core documents explicitly protect freedom of expression "regardless of frontiers," a phrase especially pertinent to the global Internet:

> *"Everyone has the right to freedom of opinion and expression; this right includes freedom to hold opinions without interference and to seek, receive and impart information and ideas through any media, and regardless of frontiers." Article 19, Universal Declaration of Human Rights.*

> *"Everyone shall have the right to freedom of expression; this right shall include freedom to seek, receive, and impart information and ideas of all kinds, regardless of frontiers, either orally, in writing or in print, in the form of art, or through any other media of his choice." Article 19, International Covenant on Civil and Political Rights.*

> *"Everyone has the right to freedom of expression. This right shall include freedom to hold opinions and to receive and impart information and ideas without interference by public authority and regardless of borders." Article 10, European Convention for the Protection of Human Rights and Fundamental Freedoms.*

No matter what the means, government restrictions on speech or access to speech of others violate basic freedom of expression protections. In addition to direct government censorship of Internet communications, or privatized censorship, freedom of speech in the Internet is threatened by diverse factors.

Blocking, filtering, and labelling techniques can restrict freedom of expression and limit access to information. Government-mandated use of blocking, filtering, and label systems violates basic international human rights protections. Global rating or labelling systems squelch the free flow of information. Efforts to force all Internet speech to be labelled or rated according to a single classification system distort the fundamental cultural diversity of the Internet and will lead to domination of one set of political or moral viewpoints. Diversity and user choice are essential: To the extent that individuals choose to employ filtering tools, it is vital that they have access to a wide variety of such tools.

"Self-regulatory" controls over Internet content, which have been promoted by some as an alternative to government regulation, ought not to place private

ISPs in the role of police officers for the Internet. With regards to content, what is being suggested in the name of "self-regulation" is not that ISPs should as a group regulate their own behaviour, but rather that they should regulate the speech of their customers. This is not true "self-regulation." The role of an Internet Service Provider is crucial for access to the Internet and because of the crucial role that they play, ISPs have been targeted by law enforcement agencies in many countries to act as content censors. While ISPs ought to provide law enforcement reasonable assistance in investigating criminal activity, confusing the role of private companies and police authorities risks substantial violation of individual civil liberties.

Privacy

The Universal Declaration, the European Convention and international human rights instruments enshrine the right to privacy. These core documents explicitly protect the privacy of correspondence and communication:

> *"No one shall be subjected to arbitrary interference with his privacy, family, home or correspondence, nor to attacks upon his honour and reputation. Everyone has the right to the protection of the law against such interference or attacks." Article 12, Universal Declaration of Human Rights*

> *"No one shall be subjected to arbitrary or unlawful interference with his privacy, family, home or correspondence, nor to unlawful attacks on his honour and reputation." Article 17, International Covenant on Civil and Political Rights*

> *"Everyone has the right to respect for his private and family life, his home and his correspondence." Article 8, European Convention for the Protection of Human Rights and Fundamental Freedoms*

Privacy is becoming increasingly important for citizens in the information society. Electronic communications can be very easily intercepted by anyone who wants to. Sending an e-mail message is thus the equivalent of sending a postcard. In the human rights arena especially, many matters discussed among NGOs are extremely confidential. Names of witnesses to human rights violations, for example, need to be kept from those who would harm them. Repressive governments commonly use their intelligence services to tap the phone communications of human rights groups and intercept their mail. It is very likely that they are also intercepting electronic mail.

How Privacy can be Protected on the Internet: Unlike phone calls and paper mail, however, electronic mail can be easily encrypted so that only those to whom it is directed can read it. This means that anyone, such as NGOs, wanting to communicate delicate information freely can do so, without having to fear the consequences of others, such as governments, accessing it.

Therefore, we believe that policies concerning cryptography should be based on the fundamental right to engage in private communication.

Unfortunately, national governments have already taken steps to detain and harass users and developers of cryptography. Nonetheless, strong cryptography is already in use by human rights advocates and its further widespread use should be affirmatively promoted. We oppose efforts that would lead to the development of communications infrastructures designed for surveillance.

Anonymity: Central to free expression and the protection of privacy is the right to express political beliefs without fear of retribution and to control the disclosure of personal identity. Protecting the right of anonymity is therefore an essential goal for the protection of personal freedoms in the online world. The right of anonymity is recognized in law and accepted by custom. It has been an integral part of the growth and development of the Internet. Some governments are working to extend techniques for anonymity. But other efforts are underway to establish mandatory identification requirements and to limit the use of techniques that protect anonymity. For example, the G-8 recently considered a proposal to require caller identification for Internet users. Some local governments have also tried to adopt legislation that would prohibit access to the Internet without the disclosure of personal identity.

Governments should not require the identification of Internet users or restrict the ability to express political beliefs on the Internet anonymously. Efforts to develop new techniques to protect anonymity and identity should be encouraged. ISPs should not establish unnecessary identification requirements for customers and should, wherever practicable, preserve the right of users to access the Internet anonymously.

7

Human Rights Approach to Poverty Reduction Strategies

BACKGROUND

Since the reforms introduced by the United Nations Secretary-General in 1997, a major task for the United Nations, and in particular for the Office of the High Commissioner for Human Rights (OHCHR), has been to integrate human rights into the whole of the Organization's work, including the overarching development goal of poverty eradication. In 2001 the United Nations Committee on Economic, Social and Cultural Rights requested the Office of the High Commissioner "to develop substantive guidelines for the integration of human rights in national poverty reduction strategies". The present document is the outcome of that request.

Its objective is to provide practitioners involved in the design and implementation of poverty reduction strategies (PRS) with operational guidelines for the adoption of a human rights approach to poverty reduction. As it is widely accepted that poverty reduction strategies must be "country-owned", the project focuses on the preparation of operational guidelines for States that are integrating human rights into their poverty reduction strategies. However, it is hoped that the guidelines will also be of use to other actors – civil society organizations, national human rights institutions, the United Nations system and other international organizations – that are committed to the eradication of poverty.

The essential idea underlying the adoption of a human rights approach to poverty reduction is that policies and institutions for poverty reduction should be based explicitly on the norms and values set out in the international law of human rights. Whether explicit or implicit, norms and values shape policies and institutions.

The human rights approach offers an explicit normative framework – that of international human rights. Underpinned by universally recognized moral values and reinforced by legal obligations, international human rights provide a compelling normative framework for the formulation of national and international policies, including poverty reduction strategies.

One reason why this framework is compelling in the context of poverty reduction is that the norms and values enshrined in it have the potential to empower the poor. It is now widely recognized that effective poverty reduction is not possible without empowerment of the poor. The human rights approach to poverty reduction is essentially about such empowerment.

The most fundamental way in which empowerment occurs is through the introduction of the concept of rights itself. Once this concept is introduced into the context of policy-making, the rationale of poverty reduction no longer derives merely from the fact that the poor have needs but also from the fact that they have rights – entitlements that give rise to legal obligations on the part of others. Poverty reduction then becomes more than charity, more than a moral obligation – it becomes a legal obligation. This recognition of the existence of legal entitlements of the poor and legal obligations of others towards them is the first step towards empowerment.

The obligations deriving from rights may be analysed by reference to the duties to respect, protect and fulfil. The duty to respect requires the duty-bearer not to breach directly or indirectly the enjoyment of any human right. The duty to protect requires the duty-bearer to take measures that prevent third parties from abusing the right. The duty to fulfil requires the duty-bearer to adopt appropriate legislative, administrative and other measures towards the full realization of human rights.

Most of the salient features of the human rights normative framework can contribute to the empowerment of the poor in one way or another. These features include the notion of accountability, the principles of universality, non-discrimination and equality, the principle of participatory decision-making processes, and recognition of the interdependence of rights. They are all essential characteristics of a human rights approach to poverty reduction.

Rights and obligations demand accountability: unless supported by a system of accountability, they may become little more than window-dressing. Accordingly, the human rights approach to poverty reduction emphasizes obligations and requires that all duty-holders, including States and intergovernmental organizations, be held to account for their conduct in relation to international human rights. While duty-holders must determine for themselves which mechanisms of accountability are most appropriate in their particular case, all mechanisms must be accessible, transparent and effective.

The twin principles of equality and non-discrimination are among the most fundamental elements of international human rights law. It follows that the international human rights normative framework has a particular preoccupation with individuals and groups who are vulnerable, marginal, disadvantaged or socially excluded. Thus, the human rights approach to poverty reduction requires that laws and institutions that foster discrimination against specific individuals and groups be eliminated and more resources devoted to areas of activity with the greatest potential to benefit the poor.

A human rights approach to poverty reduction also requires active and informed participation by the poor in the formulation, implementation and monitoring of poverty reduction strategies. The international human rights normative framework includes the right to take part in the conduct of public affairs. This is a crucial and complex human right that is inextricably linked to fundamental democratic principles. A democratic social order based on constitutionalism and free and fair elections is an essential prerequisite for enjoyment of this right. However, effective participation by the poor requires more than a functioning democracy. It calls for specific mechanisms and detailed arrangements at different levels of decision-making that help to overcome the impediments that the poor, and marginalized groups in general, face in playing an effective part in the life of the community.

The international human rights framework recognizes the interdependence of rights – the fact that the enjoyment of some rights may be dependent on or contribute to the enjoyment of others. For example, if the poor are to enjoy the right to participate in poverty reduction strategies, they must be free to organize without restriction (right of association), to meet without impediment (right of assembly), and to say what they want without intimidation (freedom of expression); they must know the relevant facts (right to information) and they must enjoy an elementary level of economic security and well-being (right to a reasonable standard of living and associated rights).

More generally, the human rights framework reflects the crucial interdependence of economic, social and cultural rights, on the one hand, and civil and political rights, on the other. Although poverty may seem to concern mainly the former category of rights, the human rights framework highlights the fact that the enjoyment of these rights may be crucially dependent on enjoyment of the latter category. A human rights approach to poverty reduction is thus holistic in nature, encompassing civil and political rights as well as economic, social and cultural rights.

Yet another feature of the human rights approach is that responsibility for poverty reduction becomes a universal obligation. While a State is primarily responsible for realizing the human rights of the people living within its jurisdiction, other States and non-State actors are also obliged to contribute to, or at the very least not to violate, human rights. This has important implications for the conduct of international affairs. It calls for an adequate flow of financial and technical assistance from the rich to the poor countries and for active efforts to establish equitable systems of multilateral trade, investment and finance that are conducive to poverty reduction.

While the human rights approach imposes an obligation on duty-holders to work towards poverty reduction, it does not make the unreasonable demand that all human rights must be realized immediately. In recognition of resource constraints, it allows, if necessary, for progressive realization of rights over a period of time and for the setting of priorities among rights in the course of progressive realization. At the same time, however, the approach imposes certain conditions on the conduct of progressive realization and on the act of prioritization so that the human rights agenda does not degenerate into mere rhetoric. For example, the human rights approach demands that minimum essential levels of all rights – or core obligations – should always be respected.

Contrary to yet another common perception, the non-fulfilment of human rights does not necessarily mean that a State is in non-compliance with its international human rights obligations. Provided that it is taking all reasonable measures towards realizing rights that are subject to progressive realization and taking immediate steps to fulfil those meant to be fulfilled without delay, the State cannot be held responsible for the fact that many rights remain unfulfilled at any point in time and hence cannot be said to be in non-compliance with its obligations. It can, however, be held responsible for not taking all measures within its power to ensure progressive realization of rights as expeditiously as possible. To help distinguish cases in which the State should be held responsible from those in which it should not is one of the functions of a proper monitoring and accountability system, which is an essential feature of the human rights approach.

ADDED VALUE PROVIDED BY THE HUMAN RIGHTS APPROACH TO POVERTY REDUCTION

Some poverty reduction strategies already have features that reflect international human rights norms. For example, the emphasis placed on civil society participation in the Poverty Reduction Strategy Paper approach advocated by the World Bank and the IMF reflects the right of individuals to take part in the conduct of public affairs, as well as the related rights of association, assembly and expression. The introduction of social safety nets resonates with the rights to a reasonable standard of living, food, housing, health protection, education and social security. Anti-poverty strategies that demand transparent budgetary and other governmental processes are consistent with the right to information, while the insistence that strategies are "country-owned" corresponds to the right of peoples to self-determination.

This congruence between features of existing poverty reduction strategies and the international human rights framework gives rise to two important points. First, the departure represented by the introduction of a rights-based approach to poverty reduction should not be overstated. Second, the application of human rights to poverty reduction reinforces some of the existing features of anti-poverty strategies. The value added by the human rights approach to poverty reduction consists both in the manner in which it departs from existing strategies and in the manner in which it reinforces them. The major elements of this process may be enumerated as follows.

By introducing the dimension of an international legal obligation, the human rights perspective adds legitimacy to the demand for making poverty reduction the primary goal of policy-making. The last decade has witnessed a move towards poverty reduction as the focus of international cooperation as well as of national policy-making. The human rights perspective adds strength to this movement by drawing attention to the fact that poverty signifies non-realization of human rights so that the adoption of a poverty reduction strategy is therefore not just desirable but obligatory on the part of States that have ratified international human rights instruments.

Recognition of the principles of equality and non-discrimination, which is a central premise of the human rights approach, helps to highlight the fact that a great deal of poverty originates from discriminatory practices – both overt and covert – at the international, national and local levels. This recognition calls for the reorientation of poverty reduction strategies from a tendency to focus on narrow economic issues towards a broader strategy that also addresses socio-cultural and political-legal institutions that sustain the structures of discrimination.

Recognition of complementarities between civil and political rights, on the one hand, and economic, social and cultural rights, on the other, adds impetus to the process of strengthening and broadening the scope of poverty reduction strategies. In particular, it helps dispel the misconception that civil and political rights and freedoms are luxuries and relevant only to relatively affluent societies. Accordingly, the human rights approach demands that measures be taken to ensure civil and political rights are integral components of poverty reduction strategies.

Unlike old-style approaches to poverty reduction, the human rights approach attaches as much importance to the processes whereby developmental goals are achieved as to the goals themselves. In particular, it emphasizes the importance of ensuring people's participation, especially participation by the poor and otherwise marginalized groups, in all aspects of decision-making. The importance of participation is being increasingly recognized. The human rights approach reinforces this recognition by drawing attention to the fact that participation is valuable not just as a means to other ends but also as a fundamental human right that should be realized for its own sake.

Making trade-offs among alternative goals in the light of social priorities and resource constraints is an integral part of any approach to policy-making. The human rights approach is no exception, but it offers the poor better protection by ruling out certain kinds of trade-offs that are potentially harmful to them. In particular, it rules out any trade-off that leads to retrogression of a human right from its existing level of realization or to non-achievement of certain minimum levels of realization that have been identified as core obligations.

Perhaps the most important source of added value in the human rights approach is the emphasis it places on the accountability of policy-makers and other actors whose actions have an impact on the rights of people. Rights imply duties, and duties demand accountability. It is therefore an intrinsic feature of the human rights approach that institutions and legal/administrative arrangements for ensuring accountability are built into any poverty reduction strategy.

In sum, the human rights approach has the potential to advance the goal of poverty reduction in a variety of ways: (a) by urging speedy adoption of a poverty reduction strategy, underpinned by human rights, as a matter of legal obligation; (b) by broadening the scope of poverty reduction strategies so as to address the structures of discrimination that generate and sustain poverty; (c) by urging the expansion of civil and political rights, which can play a crucial instrumental role in advancing the cause of poverty reduction; (d) by confirming that economic, social and cultural rights are binding international human rights, not just programmatic aspirations; (e) by adding legitimacy to the demand for ensuring meaningful participation of the poor in decision-making processes; (f) by cautioning against retrogression and non-fulfilment of minimum core obligations in the name of making trade-offs; and (g) by creating and strengthening the institutions through which policy-makers can be held accountable for their actions.

The word "guideline" is understood differently by different people and institutions. The set of Guidelines presented here should be seen as elaborating and clarifying certain principles that should guide the process of formulating, implementing and monitoring a poverty reduction strategy if it is to be consistent with a human rights approach. The expectation is that once the principles are understood, the actors involved in poverty reduction will be able to implement them in practice, keeping in view the specificities of their own context. In other words, this is not a manual that purports to teach in a mechanical, comprehensive and directive way how to implement human rights in the context of poverty reduction.

While the principles of the rights-based approach to poverty reduction enunciated in these Guidelines are broadly applicable to both rich and poor countries, the primary focus is on poverty in poorer countries. In part, this is in recognition of the obvious fact that poverty is a much more serious problem in these countries. But it is also partly because poverty in rich countries has special features that need to be addressed separately. Perhaps these Guidelines could be expanded in the future to address those issues. The Guidelines presented here do not address all aspects of human rights with equal emphasis, because they are formulated for the specific context of poverty reduction, which is only a part of the broader human rights agenda. The rationale of the choice of rights, of the relative emphasis placed on different rights, and of the targets, indicators and strategies to be found in these Guidelines should be understood in this context.

The choice of human rights to be addressed in these Guidelines was based on a judgement as to which rights are most relevant to the context of poverty. This judgement was in turn guided by the understanding that human rights can be relevant to poverty in different ways. [5] Of special significance in the context of poverty reduction are rights that have either constitutive or instrumental relevance.

The idea of constitutive relevance of human rights derives from our understanding of what it means to be poor. According to a widely accepted view, a poor person is one who is deprived of basic capabilities – such as the capability to be free form hunger, to live in good health, to be literate, and so on. In the language of rights, one may say that a poor person is one for whom a number of human rights remain unfulfilled – such as the rights to food, health, education and so on. Such rights have constitutive relevance for poverty if a person's lack of command over economic resources plays a role in causing their non-realization. Some human rights are such that their fulfilment will help realize other human rights that have constitutive relevance for poverty. For example, if the right to work is realized, it will help realize the right to food. Such rights can be said to have instrumental relevance for poverty. The same human right may, of course, have both constitutive and instrumental relevance. These Guidelines address the rights that are considered to be most relevant to poverty – on either constitutive or instrumental grounds or on both.

The Guidelines are divided into three sections. Section I sets out the basic principles of a human rights approach that should inform the process of formulating a poverty reduction strategy. Section II sets out the human rights approach to determining the content of a poverty reduction strategy. It identifies, for each of the rights relevant to poverty reduction, the major elements of a strategy for realizing that right. Here the rights are classified under two broad headings – national and international – depending on the level at which action will have to be taken. Section III explains how the human rights approach can guide the monitoring and accountability aspects of poverty reduction strategies. Like the principles discussed in Section I, accountability is also one of the basic principles of a human rights approach to poverty reduction. But because of its special significance in the context of human rights and poverty reduction, it is singled out for discussion in a separate section.

In Section II of the Guidelines, the discussion of each right is structured around four parts. Part A outlines the relevance of the right in the context of poverty. Part B sketches the scope or content of the right as set out in international human rights instruments. For ease of reference, boxes reproduce some of the most relevant international human rights provisions; they also refer to recent world conferences, as well as the most relevant General Comments or Recommendations adopted by United Nations human rights Treaty-bodies. (General Comments and Recommendations offer guidance to State parties on the contours and content of human rights and the measures that might be taken to better secure their implementation.)

Part C identifies key targets in relation to each right and lists, for each target, some indicators that will help assess the extent to which these targets are being achieved over time. Part D sets out key features of a strategy for achieving the specified targets. Some brief comments are in order regarding the targets, indicators and strategies identified in these Guidelines.

For each right, the targets were derived from the scope of the right as set out in international human rights law (and discussed in Part B under each right).

The choice of targets was guided by the following question: given the scope of the right, what are the major targets whose fulfilment would ensure realization of that particular right for the poor? Achieving these targets, for all the rights that are relevant to poverty, should be the ultimate goal of the human rights approach to poverty reduction. Many of the targets draw upon, and are similar to, those set out in the Millennium Development Goals (MDGs) adopted by the United Nations General Assembly in September 2000.

These targets may not all be achievable immediately – they may be subject to progressive realization. Nevertheless, a human rights approach to poverty reduction requires States to move as expeditiously as possible towards fulfilment of the targets. In the light of progressive realization, States should identify appropriate indicators, in relation to which they set ambitious but realistic benchmarks (i.e. intermediate targets) corresponding to each ultimate target, so that the rate of progress can be monitored and, if progress is slow, corrective action taken. Thus, indicators measure progress towards both intermediate and ultimate targets.

Several points relating to indicators are worth noting. First, the construction of human rights indicators is an ongoing enterprise and this document does not claim to have entirely resolved the matter. Nonetheless, an attempt has been made to derive from the existing literature, including that on Millennium Development Goals, a set of indicators that seem most appropriate for the targets in question, keeping in view the context of poverty. It is important to bear this poverty context in mind when inquiring why certain targets and indicators have been chosen, and not others.

Second, even in the context of poverty, the proposed list of indicators is by no means exhaustive. At the same time, it is not expected that all the indicators proposed here are applicable to all countries at all times. The list is intended for reference only. Each country must decide for itself which indicators are most appropriate for its specific circumstances.

Third, while using the indicators, it should be borne in mind that the objective is to indicate the conditions of the poor, and of specific disadvantaged groups among the poor, and not the average condition of the population as a whole. Some indicators are such that even if they refer to the overall population, any improvement or deterioration indicated by them will basically reflect changes in the condition of the poor – for example, the proportion of all children with access to primary education, or the proportion of homeless people in the country. Other indicators do not have this property – for example, the rate of specific crimes against women. These will have to be disaggregated to reflect the condition of the poor and of specially disadvantaged groups among them, e.g. poor women, minorities, indigenous peoples and so on. Exactly what type of disaggregation is appropriate will depend on the nature of the target in question and the particular circumstances of the country.

Fourth, most of the indicators proposed in these Guidelines have a striking resemblance to the standard indicators of socio-economic progress. So this raises the question, in what sense can they be characterised as human rights indicators? In brief, a human rights indicator derives from, reflects and is designed to monitor realisation or otherwise of a specific human rights norm, usually with a view to holding a duty-bearer to account. Thus, because there is overlap between human rights targets and socio-economic progress, there is likely to be a resemblance between human rights indicators and the standard indicators of socio-economic progress - although it should be observed that some human rights indicators, especially those relating to civil and political rights, do not usually figure in measures of socio-economic progress. Essentially, what tends to distinguish a human rights indicator from a standard disaggregated indicator of socio-economic progress is less its substance than (i) its explicit derivation from a human rights norm and (ii) the purpose to which it is put viz human rights monitoring with a view to holding duty-bearers to account.

Some remarks should be made to clarify the strategies proposed in connection with each right. First, the intention is not to offer a complete strategy for implementing the right in question. A strategy is offered only to the extent that it is relevant to poverty reduction. Even in the context of poverty, the idea is not to formulate comprehensive and detailed prescriptions, but to identify some broad principles that follow from human rights concerns. The details are to be worked out by the people engaged in preparing poverty reduction strategies, and these details are bound to vary depending on the context.

Second, for each right the proposed strategy consists of two distinct types of recommendation. First, there are recommendations that follow directly and clearly from the explicit human rights laws and their elaboration by the UN human rights Treaty-bodies. Second, there are recommendations that may not be explicitly stated in any treaties or human rights jurisprudence, but are considered to be consistent with the spirit and intention of the law. In general, a prescriptive tone is adopted here for the first set of recommendations, while the second is presented rather as a set of suggestions. It should be understood, however, that when the prescriptive tone is used it is only because the recommendations are in the nature of legal obligations. Moreover, these obligations have not been imposed from outside; they have been voluntarily incurred by States with a view to upholding the dignity and freedom of their people. The prescriptive language is intended merely to remind States of what they have voluntarily undertaken to do.

Third, even taking the prescriptive and suggestive elements together, the proposed strategy for any particular right cannot be regarded as comprehensive in any sense. This is partly because of the interdependence of human rights discussed earlier. Thus, the strategy proposed for implementing the right to food will not succeed for everyone unless progress is made in realizing the right to work, because most people who are not directly involved in food production have to work to purchase food from the market. Furthermore, successful implementation of any right will depend on the institutions for participation, monitoring and accountability – issues that are common to all rights and have therefore been discussed in separate sections instead of being repeated for each right. Thus, the strategy proposed for each right has to be seen as part of a comprehensive approach rather than as being adequate in isolation.

Fourth, a comprehensive approach does not prevent a State from prioritizing among rights. It is not expected that any country will find it feasible to implement all the recommendations with respect to all the rights simultaneously and immediately. Each State is free to set its own priorities and take up some of the recommendations for immediate implementation, while deferring others, so long as it abides by the principles of progressive realization. In choosing priorities, however, each State should be alert to the interdependence of human rights so as to maximize the effectiveness of its policies.

These Guidelines address large, complex, multisectoral issues of far-reaching importance. Clearly, this document represents one stage in a long-term project. While the Guidelines draw on some practical experience, they should be subjected to a piloting process and revised accordingly. In this way, they can be refined and become increasingly operational.

United Nations specialized agencies, programmes and other bodies have a wealth of material and expertise bearing upon the issues addressed in this document. Thus, they are in an excellent position to help States formulate, implement, monitor and evaluate strategies, policies, programmes and projects that reflect these Guidelines. It is to be hoped that the various United Nations agencies and programmes will coordinate closely with a view to assisting States that may wish to use the Guidelines as a way of enhancing their poverty reduction strategies.

TARGETING MEASURES USED IN ANTI-POVERTY PROGRAMS

There exist a large number of small and big PTPs in India, channelled through different ministries of the central government, and with different modalities of implementation. Some of these schemes are implemented by the state governments while others have larger proportion of funds flowing directly from the centre to district administrations. Obviously, a comprehensive review of each and every PTP would be neither feasible nor desirable. The discussion below therefore centres on a select subset of the PTPs, based upon their relative quantitative importance, availability of information relating to their implementation monitoring and evaluation, and relevance to the objectives of the present analysis, based upon targeting design and effectiveness of the scheme. The schemes chosen include examples of pure income transfer, food for work, self-employment and infrastructure generation.

First, however, a brief discussion of targeting mechanism in the Indian context is useful. A widely used categorization of targeting mechanisms that can be used to classify PTPs is as listed below.

- Self-targeting:: Such schemes rely on differential incentives of agents in tackling the problem of asymmetric information between the principal (the government providing poverty relief) and the agents (households or individuals affected by the government schemes). The design of the schemes has the objective of making the scheme worth participating in only for those who are poor, not for others.
- Activity targeting: Relies on "broad" targeting, primarily through subsidized provision of goods and services whose benefit incidence will be progressive, i.e., falling largely on those poor rather than better-off. Examples typically include primary education, provision of primary health care and basic health services in rural areas, and broadly targeted subsidies for irrigation, power and fertilizers.
- Location targeting: Based on geographical distribution of poverty, seeking to target interventions in geographic areas with high concentration of the poor.
- Indicator targeting: Relies on non-income indicators that are meant to be correlated with poverty. These can include lack of or size of ownership of land, form of dwelling, social status, gender of head of household, etc.

Asymmetric information between the government, seeking to provide transfers to the poor, and individuals or households in the economy who can legitimately or otherwise seek these transfers, is the *raison de étre* of targeting.

The underlying rationale of these targeting mechanisms is that administrative and other costs of identifying those who are poor are high, potentially reducing the resources that would be transferred to the poor under the scheme. Targeting mechanisms are a contractual/program-design innovation in response to the information asymmetry and the high costs of overcoming the information barrier. However, this framework is implicitly less than comprehensive in approach, in the sense of focusing only on one scheme at a time. In a context where the principal (in a principal-agent context) has several schemes in operation, the administrative costs *per scheme* (of overcoming information asymmetry) can get diluted substantially, thereby vitiating the need for indirect targeting mechanisms for any specific scheme.

Put alternatively, the issue of whether or not the administrative costs of identifying the poor are undertaken by the government usually does not depend on any specific scheme. In an inter-temporal context, where the government does not know what specific schemes it may want to implement in near future, "tagging the poor" – Administrative Identification – may provide externalities in terms of greater choices of schemes and their designs.

This is an important issue, as shown by the Indian experience where a large number of government poverty-targeted schemes rely on "Administrative Identification" (AI) to select beneficiaries. The most common criterion used in government schemes is that beneficiaries should be households below the poverty line (or BPL households).

Other criteria, such as focusing on SCs/STs (which per se would represent indicator targeting in the Indian context) are overlaid on the BPL status. As mentioned above, it may be argued that with an aggregate annual budget on CSS schemes exceeding Rs. 250 billion, it may be worthwhile for the government to undertake AI to better target the poor. Indeed, analytically it is perhaps more pertinent to ask why other targeting mechanisms should exist at all once AI has been undertaken.

Use of indirect targeting mechanisms in conjunction with AI may reflect in part the recognition that implementation of AI may be imperfect due to various reasons. In particular, the process itself may suffer from high Type I and Type II errors, as discussed below, resulting in exclusion of many poor and inclusion of many non-poor. In addition, the frequency of identification is necessarily spread apart in time, which would make it impossible to differentiate between transient and chronic poverty, (e.g., to differentiate the needy seeking food for work in face of natural calamity).

Administrative Identification: Tagging BPL (Below Poverty Line) Families

Since most PTPs currently in existence directly or indirectly rely on administrative classification of households into BPL and APL (Above Poverty Line), it is useful to briefly explain how this identification is undertaken. The exercise is intimately related to government efforts to provide food security to the population through the Public Distribution System (PDS). The PDS, in its earlier forms, dates back to almost fifty years ago and was a general entitlement scheme with universal coverage until 1992. It provided rationed quantity of basic food (rice, wheat, sugar, edible oils) and some essential non-food items (kerosene oil and coal) at prices substantially below market prices. The central government was responsible for procuring, storing and transporting the PDS commodities up to central warehouses in each state/union territory, while the state government was responsible for distribution within the state.

While the universal coverage of PDS continued, the government introduced two major changes, the first in 1992, in the form of the Revamped PDS (RPDS) and, subsequently, in 1997 as the Targeted PDS (TPDS), both innovations targeted at poor households. The RPDS relied on geographical targeting, being introduced with universal coverage in only 1775 blocks in poor areas – mainly tribal and hilly, drought prone and remotely located areas. The TPDS, on the other hand, was implemented in all areas but was open only to those identified as BPL. Along with the introduction of the TPDS, the price differential between PDS shops and open market was almost eliminated, effectively providing subsidized food only to BPL families.

At the core of the TPDS was division of the entire population into BPL and APL categories, based on the poverty line defined by the Planning Commission of India for different states for 1993-94. Multiple criteria were adopted for classification of BPL households, which in addition to income also included qualitative parameters like household occupation, housing conditions, number of earners, land operated or owned, live-stock, and ownership of durables such as TV, refrigerator, motor cycle/scooter, three wheelers, tractors, power tillers, combined threshers, etc. The responsibility for undertaking surveys and identifying the poor was with the state governments. However, the total number of BPL families in each state was capped somewhat arbitrarily at state-level estimates of the poor made by the Planning Commission using data for 1993-94, adjusted for growth in population in the interim.

Bibliography

Barlow K: *Ecology of Food and Nutrition*, Oceania, New Guinea, 1984.

Benjamin, J.: *Socio-religious Status of Girl Child in India*, New Delhi, Ashish Publishing, 1991.

Bhat, T.P. : *India and China : Trade Complementarities and Competitiveness*, Bookwell, Delhi, 2008.

Brems, Eva: *Human Rights: Universality and Diversity*. The Hague: Martinus Nijhoff Publishers, 2001.

Brems, Eva: *Human Rights: Universality and Diversity*. The Hague: Martinus Nijhoff Publishers, 2001.

Calman, J.: *Toward Empowerment: Women and Politics in India*, Colorado, Westview Press, 1992.

Cassese, Antonio: *Human Rights in a Changing World*. Philadelphia: Temple University Press, 1991.

Cernea, Michael, and McDowell, Christopher : *Risks and Reconstruction: Experiences of Resettlers and Refugees*. Oxford: Berghahn Books, 2000.

Claude, Inis L.: *National Minorities: An International Problem*. Cambridge, Massachusetts: Harvard University Press, 1955.

Daubenmire, F.: *Plants and Environment*, New York, Wiley, 1947.

Davis, Michael C.: *Chinese Values and Human Rights*, Hong Kong: Oxford University Press, 1995.

Deepak, B R : *India and China : 1904-2004 : A Century of Peace and Conflict*, Manak, Delhi, 2005.

Dowie, Mark: *Conservation Refugees*. Orion Magazine Nov. & Dec. 2005.

El-Hinnawi, E.: *Environmental Refugees*. Nairobi: United Nations Environment Programme, 1985.

Forbes, Anna: *Kids with AIDS*, New York, Rosen Pub. Group's Powerkids Press, 1996.

Gordon, L.: *Women and Health in America: Historical Readings*, Wisconsin, University of Wisconsin Press, 1984.

Gustafson, Carie and Peter Juviler: *Religion and Human Rights: Competing Claims?* New York: M.E. Sharpe, 1998.

Hinnawi, E.E. *Environmental Refugees*. New York: United Nations Development Program, 1985.

Hobbs, B. and Roberts, D.: *Food Poisoning and Food Hygiene*, Edward Arnold Ltd, London, 1987.

James E. : *Tibetan Treasury of Aphoristic Jewels*, Bloomington, IN: Indiana University Press, 1968.

Jane McAdam *Complementary Protection*, Oxford: Oxford University Press, 2007

Janice A. Jackson : *Tibetan Thangka Painting: Methods and Materials*, London: Serindia Publ., 1984.

Kimlycka, Will: *The Rights of Minority Cultures*. Oxford and New York: Oxford University Press, 1995.

Kimlycka, Will: *The Rights of Minority Cultures*. Oxford and New York: Oxford University Press, 1995.

Leslie, J.: *Toward Empowerment: Women and Politics in India*, Colorado, Westview Press, 1992.

Medawar, J.: *The History of Family Planning in Britain*, England, Penguin, 1971.

Mittal, Anuradha and Peter Rosset: *America Needs Human Rights*. Oakland, California: Food First Books, 1999.

Musgrave, Thomas D.: *Self-Determination and National Minorities*. Oxford: Clarendon Press, 1997.

Nicholas Steiner. *Problems of Protection*. Routledge. 2003.

Peter Willetts: *Pressure Groups in the Global System*, London, Francis Pinter, 1982

Pollock NJ: *The Concept of Food in Pacific Society: A Fijian Example*, Oceania, Fiji, 1985.

Reynard, A.: *Human Rights in Prisons*, Strasbourg: Council of Europe, 1986.

Rietman, Ed.: *Molecular Engineering of Nanosystems*, New York, Springer, 2001.

Ronald M. Davidson : *Tibetan Buddhism: Reason and Revelation*, Albany, NY: State University of New York Press, 1992.

Ross, J.A.: *Family Planning and Child Survival Programs as Assessed in 1991*, New York, Population Council, 1992.

Index

L

N

P

R

T

U

❑❑❑